Humanity in a Creative Universe

Humanity in a
Creative Universe

———•———

STUART A. KAUFFMAN

OXFORD
UNIVERSITY PRESS

OXFORD
UNIVERSITY PRESS

Oxford University Press is a department of the University of Oxford. It furthers
the University's objective of excellence in research, scholarship, and education
by publishing worldwide. Oxford is a registered trade mark of Oxford University
Press in the UK and certain other countries.

Published in the United States of America by Oxford University Press
198 Madison Avenue, New York, NY 10016, United States of America.

Library of Congress Cataloging-in-Publication Data
Names: Kauffman, Stuart A.
Title: Humanity in a creative universe / Stuart A. Kauffman.
Description: New York, NY : Oxford University Press, 2016. | Includes bibliographical references.
Identifiers: LCCN 2015030651 | ISBN 978-0-19-939045-8
Subjects: LCSH: Evolutionary developmental biology. | Evolution (Biology)
Classification: LCC QH492 .K38 2016 | DDC 576.8—dc23 LC record available at
http://lccn.loc.gov/2015030651

To Elizabeth Kauffman, In Memoriam, Katherine P. Kauffman, Ethan Kauffman, John Beebe.

Contents

PART III: *Who Are We Knowing, Doing, Living Humans
in a Creative Universe in Late Modernity?*

Acknowledgments

AT AGE SEVENTY-FIVE, I have been a scientist for some time. I have so many to thank for inspiration and friendship over so many years, mentors, colleagues, and friends, who have shaped me and this book. In no special order I thank my mentors, Hal Waddington, Warren McCulloch, and Jack Cowan, and lifelong colleague and friend, Brian Goodwin. I thank Leon Glass, Bill Macready, Ricard Solé, Eors Szathamary, Gunter von Kiedrowski, Max Aldana, Ilya Shmulevich, Sui Huang, Dennis Salahub, Leroy Hood, Philip Anderson, Kenneth Arrow, Colin Hill, Mike Brown, Nancy Kopell, Stephen Grossberg, Wojeich Zurek, Kalevi Kull, Terry Deacon, Luigi Luisi, David Daemer, Kepa Mirazo, and many others. This book owes special thanks to Giuseppe Longo, Gabor Vattay, Lee Smolin, Stanley Bates, Richard Melmon, and my wife, Katherine P. Kauffman. The weft of this book reflects so much I have learned from all the above. Undoubtedly the mistakes and ideas that may not "even be wrong" are mine alone. I thank Oxford University Press and Jeremy Lewis, my editor.

Stuart Kauffman
Crane Island, Puget Sound
July 10, 2015

Prologue

RECLAIMING ENCHANTMENT

Science alone will paint no metaphor of the face of God.
ALAN BEEM, age 26

THE OVERARCHING AIM of this book is civilizational, far beyond the science and philosophy about which I write. Human culture and the ways we humans are in the world have evolved profoundly since we were hunter-gatherers and Cro Magnon painted aurochs on the cave walls of Southern France and Northern Spain and then followed the great herds north as the last Ice Age receded. With agriculture, the accumulation of wealth became possible, and the great early civilizations of Egypt, Mesopotamia, and China arose. With the accumulation of wealth arose vast differences in wealth and power for the first time in human history. Cuneiform writing was invented in the West at about that time, as were other forms. But it is the Axial Age, more than a thousand years later, that must rivet us. In the middle of the first millennium B.C.E., across much of the world, what we now call our "classics" arose, and with them a new phase in the becoming of culture and human mind: the Hebrew prophets, Amos, Isaiah, and Jeremiah; the texts of Plato and Aristotle; the Analects of Confucius; the Daodejing; the Bhagavad-Gita; and the teachings of the Buddha in the Pali Canon. The "Axial Age" was named by Karl Jaspers shortly after World War II, who described this period as a fulcrum, an Axis, around which, across much of the globe, our sense of ourselves changed. In brief, early "religions" were about our welfare, good hunting, and good weather, appeasing and praying to local gods for intervention. With the Axial Age, for the first time, humans began to seek something "higher," the ethical for Confucius, the Dao, Buddha and Enlightenment, and Plato: We seek the Good, the True, and the Beautiful. Plato, in the allegory of the cave, never defines "The Good." It is as the sun, seen when we turn from facing the shadows cast on the cave wall, to look at it, but staring can blind us.

In diverse ways, sum Bellah, Joas, and other contributors to this book (Bellah and Joas 2012), humans sought transcendence, thus seeing themselves beyond the "human self and group welfare" of early religions.

In the West, this became the Hebraic-Hellenic roots of our common civilization. The Greeks sought universals, the ancient Jews the specificity of their lives with their God and history.

How we see the world profoundly shapes how we see ourselves, and with it, what we are and become. The Ancient world of Greco-Roman thought transformed at least with Constantine to Christianity in the West and eight or so centuries of the Church's view of God and Man and our humanity. The discovery of an intact text of *De Rerum Natura* by Lucretius in the 1300s unleashed a rediscovery of the Greco-Roman world and within a century, the Renaissance, surging beyond the Church's view of our humanity to give forth Da Vinci's *Mona Lisa* and Michelangelo's *David*. With these, came a soaring of our individual humanity. Two centuries later in the West, Descartes unleashed the scientific objective world view, culminating somewhat more than a century later with Newton and the creation of classical physics. Weber has written that with Newton we became disenchanted and entered "modernity." He was right, from Newton to the Enlightenment, the Age of Reason, where all could be, in principle, known and mastered, mystery gone, to the Industrial Revolution, to "modernity" and the diversities of "postmodernity."

The transcendence of the Axial Age is largely lost in the struggle between the two Magesteria of Stephen Jay Gould, religion and science. Science has largely won. Our spirituality is submerged.

To borrow the superb phrase of Thomas Cahill, historian, we are at a "hinge of history." Our thirty or more civilizations across the globe are weaving together, in part driven by globalization of commerce and of communication. For the first time in history we have the technological means to assure an adequate standard of living for all, or most of us, despite grotesque inequality in current wealth distribution. We live on a finite planet we despoil in the name of forever economic growth. Yet we hunger, and read the Romantic poets seeking re-enchantment. In addition to Keats, try Dylan Thomas if you have not tasted.

Are we poised, as some of the authors in Bellah and Joas's book argue, for a new Axial Age? If at a "hinge of history," what will we become? Can we see ourselves anew in the world? And as a consequence become anew?

The fullness of our human lives in a creative universe and the woven civilizations across the globe that we co-create is therefore the central issue of this book. With this, I hope to be at least one voice that may unleash an unprestatable becoming of a new Axial Age. I hope to be at least one voice that may help unleash us where we are fettered by lingering good science that blinkers us, from Newton to Darwin to Adam Smith, to John Locke, superb all, whose work laid the foundations of modernity and postmodernity in its diverse guises and sequelae, and whose work constitutes what I will call the "mythic structure" of modernity. I believe we need a new mythic framework. The aim of this book is to seek such a new mythic framework that can guide us, even as we co-create what we cannot prestate and then flow into the very Adjacent Possibilities we helped to make.

We "become." We have not known that we become into the very possibilities we unknowingly create, beyond what can be known even to be possible, as thought in the Enlightenment, the Age of Reason, where all could, in principle, be known and mastered. No, the world is not so.

If we ask: "Does whatever modernity or postmodernity may be serve our humanity?" Even with current democracy where it flourishes, and whatever the material abundance we have achieved, if grossly maldistributed, I believe many of us think the answer is "No." As Gordon Brown, then Prime Minister of the United Kingdom, said in a speech in Strasbourg, "We are reduced to price tags."

The central aim of this book is truly civilizational. My hope is that we rethink ourselves and our world and feel invited to co-create and evolve toward a new interwoven civilization beyond our own.

We will not rush to do so: the hold of the present is far too great. But if we think together and see ourselves and reality differently, we may come to be in the world in a different way.

Our current "mythic structure" derives from physics, biology, and economics: Newton, Darwin, Smith and Locke. I borrow in part from Andreas Weber, who thinks in terms of "enlivenment." With Newton, we adopted the objective view of the world, to be known and mastered. That world of classical physics is entirely entailed. Nothing emergent, not entailed, can arise. Magic is gone. Worse, we literally "lost our minds," our subjective pole, the legitimacy of our conscious experience and free will. Without these, we are not human. We also lost the central sense of ourselves as alive in a living biosphere, which we are *of*, not above. Nature is not ours to wrest our due.

From Newton we achieved and are now trapped by the view that there is a "theory of everything," reductive materialism, whose laws will "govern" and logically "entail" all that can or does become since the Big Bang, but are themselves somehow "outside of the universe." This is the Pythagorean dream of a mathematizable world. I aim to show that this view is surely false for the living world and perhaps aspects of the abiotic world. But with Newton we have come to believe we live in a world that is knowable, hence "solvable." We can optimize, find the best solution, but typically do so only to prestated problems and opportunities. Driven by this faith, technology and scientism overwhelm us. This is not real life for the evolution of the biosphere, our economy, culture, and history, where we often cannot know and cannot even prestate the opportunities and harms for becoming that will arise. Not only do we not know what *will* happen, we often do not even know what *can* happen. If we cannot prestate what *can* happen, we cannot know what can happen and thus cannot reason about it. But we must live forward anyway, and reason, the highest human virtue of our beloved Enlightenment, the Age of Reason, *partially* fails us. We become, partially beyond knowing. We, life in general, and perhaps aspects of the abiotic universe, flow into the very possibilities our status nascendi becoming creates. We "become" into an "Adjacent Possible," are "sucked into" the very possibilities we, often unknowingly, ourselves create.

Weber and Larry Arnhart rightly point beyond Newton to two of the other giants of science who shaped us and modernity and its mythic structure: Darwin and Smith. Darwin gives us evolution via natural selection, the appearance of design without a designer, hence the possibility that society can self-organize from within, not requiring an outside authority such as God for its structure. But Darwin's selection was initially too narrowly understood as "survival of the fittest," suggesting natural selection acting only at the level of the individual and its progeny, hence a competitive race. Then a kind of evolutionary "selfishness," the selection of the fitter individual with more progeny, selfishly organized the biosphere. Smith gives us the selfish individual and the "Invisible hand" where the selfish interests of the shoemaker and candlestick maker, each acting only for herself, benefit all. In this framework is the rise of modern liberalism, the view that rationality and self-organization, well framed by reason, will lead to a well-wrought society from within. The same views lead to capitalism.

We are children of the political ideals of the Enlightenment, including Locke, friend of Newton, from whom we got the idea of government as a balance of forces, in the United States, the executive branch, legislative branch, and judicial branch as mutual counterweights. We wanted no king ruling by Divine Right. Locke gives us the idea of personal liberty and property ownership, essential to Smith's invisible hand. "Life, liberty, and the pursuit of happiness," wrote Jefferson. Yes, with Locke we achieved the ideal of a republic that balances equality with our drive for power, seen in many civilizations since the agricultural revolution allowed the accumulation of massive stored wealth in the first civilizations in Mesopotamia and Egypt.

With Newton, Smith, and Darwin, Locke gives us our political framework, the fourth part of our mythic structure.

This book has three parts. The first part shows that the becoming of life, and perhaps even aspects of the abiotic universe, are beyond foundational entailing "laws," the Pythagorean dream, the dream of the "theory of everything," reductive materialism, of enormous success, and revered, but whose hegemony I seek to demolish. I hope to show you that the becoming of the biosphere is beyond law. If so, reductive materialism as a whole must fail, for the biosphere, lawless, is part of the universe and cannot be governed by a final theory. Furthermore, the becoming of the biosphere is more mysterious than we have thought, for it demonstrably creates the very possibilities into which it becomes, and it does so without selection's "acting" to achieve those possibilities. Furthermore, we must think, and explain in ways Newton did not teach us, for we cannot deduce and test from the missing laws that what we predict will happen. Not everything that is real, such as giraffes, can be predicted or known ahead of time. We really cannot always *know* even what *can* happen. Part I shows us that the becoming of life, of civilization, perhaps, even aspects of the universe, may be beyond foundational governing law and beyond knowing ahead of time. We are free to become in ways we do not yet know, but always as enabled by constraints that do not cause, but enable what, often unprestatably, next can become, status nascendi. This becoming is radically emergent. We are of it. Max Weber did say that with Newton we became disenchanted and entered modernity. Here is emergent magic to re-enchant us.

We have lost the subjective pole, our consciousness and free will able to act upon the world, to Newton and objective third-person science, and with that loss, lost our humanity. Part II is an attempt to regain that subjective pole. It is a long and speculative, partially testable, discussion based on quantum mechanics. Very recent and very tentative evidence supports aspects of the view I will discuss. It leads toward a vastly co-creative universe in which Mind, conscious and free willed, participates in how and what the actual universe becomes from electrons to us. What I shall propose is not unique: Roger Penrose in *The Emperor's New Mind* and *Shadows of the Mind* (1989, 1994) precedes me with rather similar ideas. Both Penrose and I lead us to a possible panpsychism, a pattern of ideas going back at least to Spinoza, Leibnitz, William James, C. S. Peirce, and others. Those earlier ideas were based on the seventeenth-century conception of matter and mind, trying to unite them. Quantum mechanics supersedes the seventeenth-century concept of matter in classical physics. With it, free will and mind become possible.

Part III hopes to build upon the freedom from Part I and hopes for our subjective pole from Part II. The central issue is what we may want? What might a new mythic structure be that can further guide us as we co-create what we cannot prestate? I discuss recent work on the origin of life, hence "what is life?," Erwin Schrodinger's famous question in his book of the same name, that helped found molecular biology. Our questions here have been limited to classical physics and chemistry. But quantum mechanics may play unexpected roles, as I will discuss. Given life and mind, how do we humans live it? And in the final chapter I brave the issues: Where are we? How did we get here? What might our new mythic structure contain that is beyond our current mythic structure? As a consequence, quo vadis, and how, when we cannot design but must grow, ever partially unknowing, what we become.

Among the themes of that new mythic structure beyond that of modernity are the following:

1. Beyond Newton and the entailing laws of physics to the emergent historical lawless becomings of life, culture, history, and perhaps aspects of the universe.
2. Beyond Smith and Darwin as support for selfishness as the best or dominant driving force in the self-organization of society. Ecosystems of mixed microbial species collaborate functionally in rich, but largely unknown ways, to flourish. So do we.
3. Beyond overreliance on reason. We live into the future not even knowing what can happen and have for thousands of years. Doing so is part of our human creativity. This creativity is ours, part of us in a creative universe, part of what must be in our mythos.
4. Centering on our emotional, phenomenological aliveness as individuals and in our social structures, where our emotions are the dominant source of motivations, and biologically and culturally evolved values, and the pursuit of happiness of which Jefferson wrote. But also the same evolved and cultural emotional values can play their role leading to hegemony, power structures, and violent fundamentalisms.
5. We rape the planet in the name of forever growth of gross national profit; we need an economic and cultural notion of "enough." Without it, given that rape, enough is enough.

6. Spirituality as central to our lives and fulfillment, from shaman, to gods, to *Re-ligio*—re-tie—with whatever faiths, to creative nature as sacred, but somehow without violence.

7. Our rapprochement with Nature, no longer ours to wrest our due, we are *of* Nature on a planet that should not be raped. Estranged from Nature? No.

8. An ever-clearer understanding of how wisely to garden the Adjacent Possibles, the opportunities we unwittingly co-create and into which we rush, are "sucked" but often cannot prestate, with possible new forms of governance beyond Locke.

9. We are creative in a creative universe, much of life and the universe is a status nascendi, a becoming. This is the center of our new view of ourselves and reality.

What will we make of all this? We cannot prestate, but our humanity is at stake.

I

Hinge of History

THE TRUE AIM of this book is civilizational. Indeed, *we* are the aim of this book. So much lies before us. We so wrongly think we know our worlds. We so wrongly believe that we can pose and answer our questions, when we often cannot even pose them adequately. We know intuitively that the becoming of our human lives, cultures, and economies is open and creative in ways not heeded by contemporary science. Yet science is our new reigning "god," derivative of Francis Bacon: "I take all knowledge to be my province. To put Nature on the rack and wrest our due." Wrest from Nature we have; lost from Nature we largely are. We know we despoil our home planet, but like the Venetians who cut down and denuded the nearby forests of the Dalmatian coast to build their ships, we wrest and wrest in the name of forever growth of "gross domestic product" on a finite planet.

And so much is wrong. In the United States, my home, power structures strangle us: banks too big to fail are bigger, fewer, and more consolidated than during the 2008 financial crash they largely caused. The financial industry is a larger fraction of the United States, and most of the First-World economy, than at any other time in history. We need capital formation, but we do not need what we now have that endangers us and robs us. In the United States, Congress, often responsible for our predicament, is controlled by lobbyists, who are often chosen from those no longer in Congress but with rich political contacts—a clear conflict of public interest. At a minimum we need to know how power structures form and how they evolve and adapt. I will try to present a new framework to think about this in later chapters, for we are not helpless. The aim is not to destroy dangerous power structures but, for the benefit of all, to help transform them.

A close friend of mine once said that if poetry is not at the center of a civilization, something is deeply amiss. Where is art in our modern world?

Perhaps it remains best seen in cinema in our wider culture. But poetry, art, music, even plays, are less among us, enamored as we are of Walmart. Poetry is metaphor, neither true nor false, but rich and evocative of our wider selves.

Facts and true statements are wonderful, rich, and enabling but also too much with us. Michelangelo's *David* is not a "fact" true or false; it is "us." Propositions and logic are wonderful and richly enabling of much of language, mathematics, and the life and technology we have created in part with it, at the price of categorizing the world. But the world is and becomes, beyond our categorizations, often beyond "scientific law," I hope to show, and we are of this creative world, beyond "sciencia." "Would you rather be Einstein or Shakespeare?" I asked myself, when I was younger. Einstein, of course. But that is not my answer now, magnificent as he was. It is Shakespeare who taught us what it is to be human. "Everyman" is more than science.

Arnold Toynbee in *A Study of History* (1934–1961) famously argued that civilizations are born, mature, and ultimately die. Typically, he argued, a spiritual rebirth in the dying age of a civilization lays the foundation for the new, unexpected, unprestatable civilization that emerges. He may overstate his case, but the rise of Christianity in the late Roman Empire bespoke a thousand years of a new church view of our humanity, a view of our humanity transformed by the Renaissance's rediscovery of antiquity unleashed by finding a full manuscript in the monastery of Fulda, of *De Rerum Natura* by Lucretius. In the Prologue, I spoke of the "mythic structure" of First-World modernity and its sequelae, derived from Newton, Darwin, Adam Smith, and Locke. *De Rerum Natura* unleashed Florence, Michelangelo, da Vinci, the Medici, and the Renaissance—an entire new view of our humanity.

And now? The Renaissance became the Enlightenment, the Industrial Revolution, modernity, and then postmodernity. Newton's laws, as we shall see, entail the becoming of the classical world, and we are mindless machines in it. Darwin gives us descent with modification and natural selection, the appearance of design without a designer, via "survival of the fittest," too easily translated into mere selfish competition, nature red in tooth and claw. Darwin, monumentally, gives us in Western science *history*; Smith gives us selfish individuals, human only in preferences and trade decisions. With Locke as well, modernity is built upon these conceptual foundations, our "mythic structure."

I have spoken to a modest number of people; we all sense something deeply deficient in our modern civilization. Is it an absence of spirituality? Partly. A greedy materialism beyond what we really need? Yes, we are riding the tiger of late capitalism, where we make our living producing, selling, and buying goods and services we often do not need on this finite planet. We are hanging onto the tail of this tiger, in a sense not knowing what else to do. "Purple plastic penguins for the poolside?" Well, I want no culture czar telling me "No." But do we truly need much of what we make and sell? No, probably not. What else might we do? What about art, poetry, adventure, joy, laughter, silliness, honor, integrity, love? What about just being ourselves flowering? But then, what is it for *us to flower*? What, then, is our humanity? We cannot see ourselves, in part blinkered by unneeded "scientism," technology and the blinding lights of apps, and ads, including those for medicines where the dangerous side effects are quietly spoken under cover of soft romantic music. Sell or die. Why? Why so excessively? Tail of the tiger, that's why. But then, Why?

We have forgotten that we are, first of all, alive, and alive in a becoming biosphere, a rich and almost unfathomable becoming, as we shall see, in ways we cannot prestate, yet is somehow coherent. Despite bursts of extinction events and the fact that 99 percent of all species are gone, the biosphere flowers on. And on and on and on, ever-becoming beyond what we can say ahead of time.

This flowering of the biosphere, more than a metaphor for human history, begins to suggest a mythic structure beyond that we live by. We need no longer choose to do so, and we can find a new mythic structure for our transformation.

We can, for the first time in history, afford a wider civilization where our material needs are largely met. This affords us a unique choice: We can afford to become what we can partly envision and partly co-create, although ever-unknowing of all we will unleash. Thus, our central issues are "What do we hope for, what is our dream, if only partially statable? How do we start?" Andreas Weber, in his *Enlivenment* (Weber 2013), dreams of us individually and together living "fully," where living fully becomes our single and transcultural dream, sustained by the new mythic structure I hope this book contributes to. Fully alive, like the becoming biosphere, it is a confused mess of becoming, collaboration, competition, coherence and its failures, ever-new, ever-unfolding, creative, co-creative, creating the very possibilities into which we almost ineluctably become. This theme of becoming into the ever-new, largely unprestatable, Adjacent Possibles we ourselves persistently create is a main theme of this book.

Part I: Beyond Reductive Materialism

Part I (Chapters 3–5) seeks to state and at least to destroy the hegemony of reductive materialism, the magnificent structure built by Descartes, Kepler, Galileo, Newton, Laplace, Einstein, Bohr, Schrodinger, and others hoping that a final theory governs and hence entails all that can and will become in the universe. But if no law entails or "governs" the becoming of the biosphere, which is part of the universe, reductive materialism must, in broad view, fail: If no law entails the becoming of the biosphere, part of the universe, then *no* law entails the becoming of the entire universe! Weinberg's dream of a final theory must fail (2014). Moreover, I hope to convince you that no law entails or governs the becoming of the economy, or the social cultural world, in part due to the conscious choice afforded by free will, which is discussed in Part II. Part I sets us free. We are not entirely entailed by governing laws. We will find a living world that is a world of unprestatable becomings, a status nascendi, of co-creation beyond our knowing beforehand. Reason will often fail us: not only do we not know what *will* happen, but we often do not know what *can* happen. If not, we cannot reason about what we do not know *can* happen.

We will find a new pattern of explanation for the living world: ever-new Actuals do not cause, but *enable* ever-new, often nonstatable, adjacent-possible opportunities into which we are "sucked." We will find ourselves beyond Newton, Darwin, Adam Smith, and Locke, thus past our First-World mythic structure. We will be forced to change that mythic structure. What a stunning opportunity: we can begin to shape the mythic structure that can guide us beyond whatever late modernity is, into what we can co-create but cannot control or prestate. We become! We are status nascendi, a generalization of improvisational comedy where none in the troupe, us, knows beforehand what we will co-create: we co-create our lives and our evolving woven global civilization.

Further, I will dare to ask if the Pythagorean dream of foundational laws, themselves outside of the universe, is forced upon us. I think not. All may be emergent and ultimately lawless at base. All may be a status nascendi. All may be emergent and without foundation in Pythagorean "law."

All this matters! We are, in fact, at a hinge of history. Our thirty or more civilizations are thrown ever-more into intimate contact, sparking opportunities to co-invent, and rage, as cultures collide. Before us, as never earlier in human history, is what we will make of ourselves and our home planet.

Demolishing the hegemony or even *all* of reductive materialism, I dream, sets us free to, well, dream and create.

Part II: Seeking Our Subjective Pole

Part II spans from Chapters 6 through 14. With Newton and classical physics, we vaulted into third-person objective science, mastered how to get to Mars, but lost our minds. We have become either zombies, and at best have no free will, or if conscious, that consciousness cannot alter the world it merely witnesses. Our consciousness is merely "epiphenomenal," with no effect on the world. Then why did our complex brains evolve?

We will never get beyond this, trapped here since Descartes' res cogitans and res extensa led via Newton to the death of res cogitans and the triumph of res extensa, thus classical physics up to general relativity. The stalemate yielding at best a witnessing, epiphenomenal consciousness and no free will is precisely due to the causal closure of classical physics. To get beyond the stalemate we must go beyond classical physics.

Consider this: Several years ago, NASA chose to design and send a rocket to Mars. The rocket landed on Mars, slightly changing its mass, hence the dynamics of at least the solar system. We all believe, and in the United States, we fund NASA, to make such decisions, design or not such craft, and carry out or not such missions. We believe we have free will to choose, could have, contrary to fact, chosen otherwise, can responsibly act, and in doing so we can change the world, even the solar system. Our legal system assumes the same, as do our political and economic systems, Smith and Locke included. But on classical physics, none of this can be true. A first way to see this is that in classical physics, given initial and boundary conditions and the laws of motion of the system, say billiard balls rolling on a fixed table, only what *actually* happens can have happened. The present could *not* have been different! But free will to choose or not demands that the present *could*, counterfactually, have been different. We are lost if we stay in the classical physics stalemate. But quantum mechanics affords a, or perhaps *the* way forward. In quantum mechanics, the outcome of a quantum "measurement" could have been different, so the present could have been different. This is a minimal ontological requirement for a free will. Only quantum mechanics offers ontological hope for our capacity to choose or not to send a rocket to Mars, altering the dynamics of the solar system. More quantum mechanics points beyond physics, to a role for the observer in what arises.

No one thought that quantum mechanics could matter in the warm wet world of biology. However, in the past number of years it has become clear that this view is wrong. Light-harvesting molecules can be quantum coherent for a thousand times longer than thought, and their quantum behavior is related to the efficiency of converting harvested light to chemical energy in photosynthesis that drives much of life (Engel et al. 2007). Bird migration in the Earth's magnetic field now seems to require quantum effects (Gane et al. 2013). Our best hope for a conscious mind (and free will) that can be part of altering the world, not merely "epiphenomenally" witnessing it, lies in quantum mechanics. We've lost our humanity to amazing Newton and objective science. Quantum mechanics can give us back our subjective pole, hence our richly evolved humanity.

Part II is, therefore, a long discussion of quantum mechanics. I discuss a new "state of matter" hovering reversibly between quantum coherent and "classical behavior," the Poised Realm, patent issued and issuing (Kauffman et al. 2014). The Poised Realm breaks the causal closure of classical physics and allows quantum and poised realm mind to have *acausal* consequences for the "meat" of the brain and body. Mind and brain and body and we together can act in the world with free will. The same ideas lead to trans-Turing systems that are quantum, Poised Realm and classical, and *not* Turing machine algorithmic. Our minds need not be algorithmic or merely epiphenomenal. However, no third-person description can yield us consciousness, as David Chalmers argues (Chalmers 1996).

To go further, I offer a new interpretation of quantum mechanics and a new analysis of what is called the quantum enigma (Rosenblum and Kuttner 1996). I am not a physicist, so be very skeptical. One of the central ideas I broach, almost demanded by quantum-coherent behavior which fails to obey Aristotle's law of the excluded middle, is that quantum-coherent behavior cannot be about Actuals that do obey that law, but it can be about ontologically *real possibles*, which do not obey the law of the excluded middle. I will propose res potentia and res extensa linked by measurement. This naming is in honor of Descartes. But Descartes offered us a failed substance dualism, res cogitans, like res extensa is "stuff." Res potentia is not "stuff" because possibles are not "stuff." My dualism is new—not prey to the devastating critiques against any substance dualism. On this view the world consists of at least a dualism, Possibles and Actuals. The postulate of ontologically real Possibles, res potentia, is in none of the other perhaps fourteen interpretations of quantum mechanics. More the postulate of ontologically real Possibles, res potentia, answers at least four deep and unexplained mysteries of quantum mechanics. First among these is the enormous mystery of "nonlocality." Here two widely separated but "quantum entangled" particles have the property that measurement of the first *instantaneously* alters the outcome of the later measurement of the second, even though no light can travel between the two locations of measurement in the time interval between measurements. Because light is the maximum but finite velocity of any causal effects, the instantaneous correlation of the two entangled measurements cannot be due to causal effects. This radical prediction, first made by Einstein, Podolsky, and Rosen in 1935 to show that quantum mechanics must be incomplete with "spooky action at a distance" has now been very well confirmed experimentally. The world, like it or not, has effects, called "nonlocal" that cannot be causal. As I will show, the postulate of ontologically real possibles seems to explain nonlocality easily. No other explanation for nonlocality seems to be at hand or accepted. Ontologically real possibilities can explain three other mysteries, as I will discuss in Part II.

But none of this gets us to consciousness or free will. Quantum mechanics and the quantum enigma (Rosenblum and Kuttner 1996) seem to afford an avenue. Bohr (1948), in the Copenhagen interpretation of quantum mechanics, and von Neumann in his 1933 masterful mathematical treatment of quantum mechanics and quantum measurement, both thought that human conscious observation could mediate measurement. Very recently, very tentative experimental evidence by Dean Radin (2012, 2013) suggests that human conscious attention, even at a distance, can alter measurement. We are not to believe Radin's results as yet. But they point one way to demonstrate that human conscious attention can alter the outcome of quantum measurement.

Further, the quantum enigma demands much the same and more. The enigma states, roughly, that we can ask a question of nature, but we could have asked a different question of nature; that is, we could do one experiment but could, free-willed, have chosen, counterfactually, to have done another experiment. Then Nature answers, and the answer *depends upon the question we free-willed chose to ask*. Jointly, we and Nature *create* Reality. Are you shocked? But we sent a rocket to Mars, and did so, we think, consciously and free-willed, but we could have done otherwise and altered the dynamics of the solar system.

To solve the enigma, I will propose that we are conscious and so are quantum variables such as electrons and protons exchanging photons measuring one another, where measurement is mediated, I claim, by consciously observing one another. I cannot see any way of showing that electrons consciously measure one other, but Radin's experiments are a first hint that we can show

how human consciousness can "mediate" measurement, perhaps even nonlocally. And there are other experimental approaches to show the same thing.
If this be true, after much evidence, then, as we shall see, physicists claim that
classical recording devices can mediate measurement. But then we either have
to "invent" another means for classical devices to mediate measurement, or
we try the idea that at base, "classical" devices are quantum and the quantum
variables consciously measure one another.

This all leads to a vast panpsychism, in which all quantum measurement
is mediated by Mind, conscious and free-willed, as part of the furniture of
the entire universe! It is the enigma all the way down. Mind is part of the
actual becoming of the universe, we altering the mass of Mars, down to any
quantum measurement and its Actual outcome! Each such measurement
alters what the world becomes. If so, if measurement is mind measuring consciously and with free will, bounded by what is called the Born rule, then
the entire becoming of the universe is *not entailed*. We are profoundly free! I
will expand my dualism, res potentia and res extensa linked by measurement
to be measurement acausally by conscious, free-willed *mind*. I am led to a
new triad: Actuals, Possibles, and Mind. Mind measures Possibles to yield,
acausally, new in the universe Actuals, which acausally yield new Possibles
for Mind to measure to again yield new Actuals yielding new Possibles for
Mind, with free-will, to measure. All is a becoming, nothing *is*, and all is
an unentailed status nascendi, entailed in detail by nothing. Penrose's view
(1989, 1994) is a cousin of these ideas.

Yes, Shakespeare: "Horatio, the World is richer than all your philoso-
phies." But pause: The Triad, like all interpretations of quantum mechanics,
is insufficient. Whence the "classical world"? We get to Mars using Newton's
classical physics laws, and gravitational lensing according to classical physics
general relativity is well established. There are diverse tries at getting from the
quantum to classical world. I shall offer a new and testable proposal based
on what is called the quantum Zeno effect, just a hope at present. The mind-
body problem requires a body. Moreover, evolution could not have happened
without a "stable-enough" classical world in which adaptations could have
accumulated. We must have a classical-enough world as well as something
like the Triad, or the different attempt of Orch Or by Penrose and Hameroff
(Penrose 1989, 1994) that also yields a consciousness and probably free will
as part of the universe, hence mind and a panpsychism similar to the Triad.
In the meantime, efforts to unite quantum mechanics with general relativity
have failed since 1927. We may find a way, or we may be lost in the wrong

forest trying to do so. I suspect the wrong forest, but I am not a physicist. Just maybe we can get a classical-enough world from the quantum Zeno effect and need not quantize gravity, a claim that is at best a hope, and must infuriate all physicists seeking a unified theory quantizing gravity that will entail all that becomes in the universe, including the unentailed evolution of the biosphere and legal proceedings, where I might or might not be found guilty and hung.

Part III: Regaining Our Fully Alive Humanity

Parts I and II hope to have set us free and have found at least one route, perhaps with Penrose (1989), who includes his work with Hameroff in Penrose (1994), in parallel, to our Subjective Pole with consciousness and sometimes responsible free will. But if these belong to quantum variables (or space-time itself in Penrose and Hammeroff), Mind is surely far different in life and humans. Consciousness and Will have evolved at the origin and with *life*. We are alive first. I discuss recent work that may be closing in on one or more ways life may have originated, perhaps conscious and free-willed from the outset, here and perhaps widely in the universe. We may succeed soon in creating life anew, classical or quantum, Poised Realm and classical. Mind and its living embodiment have evolved on Earth for 3.7 billion years. What is it like to be a bat? wrote Thomas Nagel (1974). Indeed, but what is it "like" to be *Escherichia coli*, the single-celled Stentor that gives signs of emotions ranging from fear to disgust and anger, hydra, a tree, whale, chimp, or us? What is the evolution of our phenomenology and its relation to our embodiment? With mind and body, with our sense of "self" alone and with others, with emotions, cognitions, intuitions, and sensations? What is it to be alive, human, and co-create our worlds? How do we live forward and choose when we often cannot know even what *can* happen? Most centrally, we are largely free, enabling by our laws and culture but not knowing the adjacent-possible opportunities that we co-create. How and for what will we dare and deign to choose and act? We can, for the first time in this hinge of history when our thirty or more civilizations weave together, afford a sustainable life for all. But what will guide us as we choose, not knowing the consequences of our choices? By the very creativity of the universe and we in it, we cannot *know* what we will co-create. Then what can guide us?

Our guide can be a new founding mythic structure that reflects our full enlivenment (Weber 2013), humanity in a creative universe, biosphere and human individual and social lives that are fully lived and keep becoming. The dream is diversity, more ways of being human as our thirty or so civilizations across the globe weave together gently enough to honor their roots and allow change to unfold gracefully. Horatio, as ever, our global woven civilization is ours to create, ever-unknowing, facing, as Kant said, the crooked timber of our humanity.

2

The Foundations

THE EMERGENCE OF REDUCTIVE MATERIALISM
AND THE LOSS OF OUR HUMANITY

MY PRIMARY GOAL in this second chapter is to lay out the overwhelming scientific framework of reductive materialism, which bespeaks most of our scientific view of the world. If we have to pick a singular figure as its dominant author, that figure is Newton. The early sociologist, Max Weber, wrote, as I noted earlier, "With Newton we became disenchanted and entered Modernity." Weber was right: classical physics, the Enlightenment with the demand for the rise of science and "down with the clerics," the Industrial Revolution and modernity with its Enlightenment heritage in political philosophy, in part the child of Adam Smith and Darwin, along with Locke, the massive technology from the Industrial Revolution, and an overarching "scientism." We remain disenchanted in many ways, and we live in a modernity or a postmodernity—whatever exactly we may mean by "modernity"—that serves us well yet does us a profound disservice. I have been a scientist for over fifty years. I love science and admire the vast achievements of modern science, conceptually, mathematically, experimentally, and technologically. But a main point of this book is that science is too much with us, our humanity vastly outstrips science, and it is our humanity and a potential transformation to a civilization that unleashes and serves that humanity as it ever becomes in ever novelty that must be our foremost purpose.

I do not seek to destroy the magnificent house that Newton built, but to end its dominating hegemony over our view of reality. After all, we can and still do get to Saturn in rockets using Newton's laws, and quantum mechanics is validated to eleven decimal places. Modern science is superb. But modern science dominates us beyond what we need.

Furthermore, modern science asserts, assumes, and demands the Pythagorean dream that there *must* be foundational laws, currently general relativity and quantum mechanics, whose laws stand "outside the universe," for in the Pythagorean dream, we may not ask, *Why these laws?* The laws just *are*. Part of what I hope to do in this book is to show us that we are not forced to adopt the Pythagorean dream, which in turn gave rise to reductive materialism. We may be foundationless and ultimately in a co-creative universe. But that radical and highly speculative view is for later. First we need to see our assumed foundations and how we got here.

I begin briefly with Aristotle and science in ancient Greece. Largely this science was not experimental. Thus, for Aristotle, it was clear that moving objects came to rest because absence of motion was natural, that they remained on the surface of the Earth because they sought a natural resting place at the center of the Earth, and that planets in known motion against the fixed stars would follow circular orbits because circles were the most perfect form. I overstate my case. Archimedes' cry, "Eureka," when he understood why some objects floated when the water they displaced was more than the weight of the object, was clearly a mixture of empirical and imaginative, creative science Einstein could have well adored. Indeed, Newton's first law was formulated by a Hellenic physicist about 150 A.D. and Lucretius' *De Rerum Natura*, whose rediscovery helped unleash the Renaissance, invented the "swerve" by moving particles to avoid the utter determinism of Democritus and his atomism, atoms (uncutables) moving in the void and colliding and rebounding deterministically. Lucretius came close to Newton's law of gravity as well: a stone falling through a cylindrical hole through the center of the Earth would not come to rest at the center of the Earth, as Aristotle taught, but oscillate back and forth in damped oscillations until it finally came to rest at the center of the Earth.

I noted earlier that Arnold Toynbee (1934–1961) argues, perhaps too strongly, that civilizations are born, mature, and die. In their death, often or typically there is some form of unexpected spiritual rebirth that forms the seeds of the new civilization. Jesus and Christianity were the seeds as the Hellenic world and Roman Empire failed. With Constantine and the adoption of Christianity by the Roman Empire, for religious and political reasons, the Middle Ages arose, a new civilization, with the Church's view of our humanity that dominated us until the Renaissance began to destroy the Church's view and a new civilization began to arise with a new view of our humanity. This view was captured by Michelangelo's treasures, including *David* and the Sistine Chapel's ceiling, God's finger reaching out and almost, but not quite, touching the outstretched finger of Adam. How perfect that the two fingers almost but not quite touch. *David*, and God and Adam's fingers nearly touching, evoke; they do not state. Art, unlike science, is not about true and false propositions; it is metaphor, neither true nor false in the sense in which "The cat is on the mat" is true or false and we take science to be true or false. Metaphor evokes. We live richly by metaphor in all our art. Art tells us, Sophocles and Shakespeare, that human aliveness evocable by metaphor is far more than "sciencia," that is, knowing. This too is a major theme of this book. We, alive, are more than we know or can know. With the Renaissance arose a new, unprestatable view of our humanity that ultimately led to modernity via Newton.

One major step in the birth of modernity was the birth of science inspired by Francis Bacon, who argued that we no longer limit ourselves to reading "the philosopher" Aristotle, but look to the book of Nature. Bacon bravely or proudly: "I take all knowledge to be my province" and "To put Nature on the rack and wrest our due!" Genesis, with the reading of Creation, made for humanity's use, not stewardship, echoes in "Wrest our due." Modernity and its overwrought technology skid erratically into the future, as promised by Bacon. We wrest our due and destroy the Earth.

Four other figures dominate the birth of Western science, all well known. Copernicus offered the hypothesis that the planets revolve around the Sun. Kepler, the last of the white magicians, started with the hope that the five planets would follow the five perfect Platonic solids in their orbits but used Tycho Brahe's data to work out that the planets follow elliptical, not circular, orbits around the Earth or Sun. And then Galileo clearly promulgated the view that nature could be described in detail by mathematical laws in his experiments with acceleration of balls rolling down incline planes whose slope "diluted" gravity, and distance covered varies as the square of the time elapsed. And then Descartes. Follow please: Early mathematics before Euclid was largely sums and accounting. With Euclid we get theorems by construction and deduction and even irrational numbers by Pythagoras. The Indians brought us zero and with them negative numbers, and the Islamic civilization brought us algebra, that is, equations such as $y = x - 7$ or $y = x^2$. Given equations with, here, two variables, y and x, Descartes had his famous dream of "analytic geometry" in which all equations would each correspond to some line locus, straight or curved in a Cartesian x, y coordinate system.

In 1640, Descartes set the framework of modern science, with his famous substance dualism, res cogitans, thinking stuff attempting to preserve our "subjective pole," and res extensa, his mechanical worldview, which hoped to be our objective worldview that would, with Newton, become classical physics. But res cogitans failed because, put simply, no one could see, then or now, how res cogitans could "act" causally or otherwise on res extensa. With this, we literally lost our minds. At the center of our current view of the world, we have lost our subjective pole to science since then. Trying to recover our subjective pole will be the central topic of Part II of this book. I think we can recover a mind that can act, but acausally, on brain and body and the world. I think we can have responsible free will. What I shall write in Part II is new science and philosophy, but testable. I am led, not as a physicist, to a new formulation of quantum mechanics. So caveat lector: not a physicist. You will see what you think of the ideas. These are issues, I hope, we can largely settle. If science, in particular classical physics, has largely stripped us of an account of our subjective pole, central to our humanity, I think perhaps we can regain a scientific account of our subjectivity, helping us regain our humanity in quantum mechanics.

But this Chapter 2 wishes to lay the foundations of our current objective scientific worldview, which has led to the hegemony of "reductive materialism." Reductive materialism, in which the entire becoming of the world since the Big Bang is "entailed" by underlying physical laws "outside the universe," also robs us of our humanity. We are "entailed." As Weber said, "With Newton, we became disenchanted and entered modernity." Descartes unleashed Newton, who created classical physics with which Laplace led to the birth of the reductive materialism whose hegemony, even necessity, I wish to challenge in Part I of this book. I wish to show that in many ways our becoming is not "entailed." If so, reductive materialism as a dream for all that is, the Pythagorean dream, also fails.

One mind, abetted by his competitor, Leibnitz, invented the differential and integral calculus, where the differential calculus is the equation for the slope of a line tangent to a Cartesian equation's curve at each point. Thus, for $y = x^2$, the equation for the "derivative" is $2x$, the tangent slope of x^2 at each point, x. $2x$ gives the "instantaneous" rate of change of y with respect to x at any point on the $y = x^2$ curve. Then Newton invented Integration: given an equation for an initial curve in a Cartesian coordinate system, say $y = x^2$, that equation is the derivative at each point on that curve for the area under that $y = x^2$ curve. The definite area under that curve can be defined by setting lower and upper limits to x.

Then Newton invented his three laws of motion, giving the forces between particles such as balls on a billiard table; when written in the differential equation form, he invented Law 1. A body continues in straight line motion unless disturbed by an outside force (contra Aristotle), $dx/dt = C$, the change in position, dx, at each tiny interval or, in the limit of $dt \to 0$, instant in time, dt is constant, C. More there is no change in the direction of motion; bodies move in straight lines, due to "mysterious Inertia." In Law 2, acceleration of motion, in velocity or change of direction, requires a force, $F = ma$; force is equal to mass times acceleration. In Law 3, every action has an equal and opposite reaction. Newton wrote the last two laws in differential equation form. Actually he used geometric arguments in the *Principia*, not differential equations, and he never took the limit $dt \to 0$, for he thought that limit violated causality. Leibnitz took the $dt \to 0$ limit, and calculus in the next centuries derives from Leibnitz.

I comment that Aristotle had four causes: material, formal, final, and efficient. The material cause of a house is bricks or wood or stone. The formal cause of a house is its design. The final cause of a house is your desire to build it, your "telos." The efficient cause of a house is the set of builders stacking bricks and the sculptor chiseling marble. Newton mathematized Aristotle's efficient cause (Rosen 1991). With the success of classical physics, soon to follow, Aristotle's other causes fell from science. We will recover them next in diverse senses.

Newton added another law, not an efficient cause law: universal gravitation, acting instantaneously, hence not an efficient cause, is an attractive force proportional to the product of the masses of the two mutually gravitating objects divided by the square of the distance between their centers of mass.

Newton's laws yield classical physics, causally closed, and an entirely "entailed" view of the becoming of reality. The conceptual framework invented by Newton is stunning for its brilliance, for its pervasive power, and for the hold it retains over our minds. Newton remains our dominant model for how to do science. "God said let there be Newton, And all was Light," wrote Alexander Pope.

What had Newton done? Consider seven billiard balls rolling on a billiard table. "Isaac," I ask, "what will happen to these rolling balls?" "Stuart, I told you what to do. Measure the present positions and momenta, that is, velocity times mass, of each of the balls, which I call the 'initial conditions.' Measure the boundaries of the table that the balls will bounce off of. These are the 'boundary conditions.' Write down in differential equation form my laws of motion, giving the forces acting between the balls as they may hit one another, and hit and bounce off the walls. Now *integrate* these differential equations and you will get, ignoring friction for the moment, the trajectories the balls will trace out on the table, forever."

With respect to the Sun and Earth, integration of Newton's universal law of gravitation yielded the amazing result that the orbit of the Earth around the Sun would be an ellipse, exactly fulfilling Kepler's three laws. Perhaps this triumph, uniting motion on Earth and in space, did more to inspire confidence in Newton's work than any other and led to the birth of classical physics.

I now ask a central question: what is it to *integrate* Newton's differential equations of motion? It is to *deduce* the consequences of the differential equations of motion for the forever trajectories of the billiard balls on the table. But "deduction" is *logical entailment*. "All men are mortal. Socrates is a man. Therefore, Socrates is a mortal." The truth of the premises entails the truth of the conclusion. The "truth" of Newton's laws of motion and universal gravitation in differential equation form, given initial and boundary conditions, entirely entails the deduced trajectories of the billiard balls on the table or Earth around the Sun.

Furthermore, Aristotle had said that there is "no prime mover" or uncaused consequence. This holds in Newton: there are no changes in motion that are not caused by forces among the particles and boundaries. At its birth, classical physics is *causally closed*. This will have enormous bearing in Part II, for the causal closure of classical physics underlies many of our problems in the philosophy of mind where we wonder how mind could "act causally" on the "meat of the brain." To see this simply, can my mind alter the motion of the billiard balls? No, we say, and we have, more or less, been frozen here since Descartes' res cogitans failed, given the explosion of his res extensa with Newton. How is res cogitans to act causally on res extensa? I hope to provide answers in Part II, which is essential for Newton, at a stroke, robbed us of our conscious mind and responsible free will altering the world.

Reductive Materialism, the Modern Form of the Pythagorean Dream

How did we get to reductive materialism? To answer this, I turn to Laplace and the birth of modern reductive materialism.

At the time of Napoleon, the great Simon Pierre Laplace imagined a "demon" who knew the positions and momenta of all the particles in the universe. Such a Laplacian demon could calculate, said Laplace, the entire future and past history of the universe. This *is* the birth of reductive materialism in its modern form. For Newton's differential equations, once integrated, *entail* trajectories of the system in its phase space, here the entire universe. Newton's laws are time reversible; hence, the past as well as the future could be calculated. This is the first clear foundation of modern reductive materialism. The dream is that there are laws "down there" that entail the *entire* becoming of the universe since its formation in what we now call the "Big Bang." As we will see, this is somewhat modified by quantum mechanics, but not in its major outlines.

This is the Pythagorean dream, in which the fundamental laws stand outside the universe and just *are*. We cannot ask, "Why these laws," because questions stop at the laws. All that becomes is entailed. Nothing not entailed can arise. There is no magic because all is entailed. Moreover, we are disenchanted and mindless.

Laplace assumed with all physicists that determinism implied predictability. He was often right but in general wrong. The basic idea is that nearby initial positions and momenta follow forever "sufficiently" nearby trajectories and hence are predictable. This is often true, but not always. Newton integrated his law of universal gravitation for the Earth in orbit around the Sun to get his famous analytical solution: the orbit is given by the analytical equation for an ellipse. But Newton worried that he might have problems with three gravitating objects, Sun, Earth, and Venus. In about 1898, the famous Henri Poincaré changed our view of the world by showing the following: (1) There is no analytical (equation) solution to the three gravitating body problem or *n* body, for *n* greater or equal to 3. (2) For the first time, Poincaré showed what is now known as "deterministic chaos." Here two "infinitely close" initial positions and momenta can follow trajectories that veer apart, becoming "exponentially" more distant with time. The exponential rate of divergence of trajectories is measured by a positive "Lyapunov exponent" giving that exponential rate. But then determinism no longer implies predictability for we cannot measure initial conditions to infinite accuracy. The weather is an example of such a chaotic system.

"Here enters randomness," "Ice entre l'aleatoire," wrote Poincaré.

Deterministic chaos is now a well-studied subset of classical physics and the theory of nonlinear differential equations. Chaos is found in many places, from the weather, as noted earlier, which is why long-term predictions of weather fail, and even in normal cardiac rhythms.

Implications of Reductive Materialism

The triumph of classical physics, beginning with Newton, Leibnitz, and Laplace, the vast development of the differential and integral calculus, its applications to ever-wider areas of physics in the seventeenth, eighteenth, and nineteenth centuries, led to profound transformations in our view of reality.

Among the most profound implications of classical physics and reductive materialism, the view held with growing confidence by physicists, and slowly much of the lay Western public was to cast severe doubt on a theistic supernatural God who acted in the universe, as averred by the Old and New Testaments. Nature, after Newton, was seen to be "governed" by deterministic laws. Then there was no room for miracles wrought by God or Jesus. The struggle between faith and reason was not new. "Batter my heart, three-person'd God," begins John Donne's magnificent holy sonnet aching for faith, "Nor ever chaste, except you ravish me." But Donne came before Newton *deduced* Kepler's elliptical orbits of the planets. God was transformed for many from the theistic God to a deistic God who wound up the universe, stood back, and let Newton's laws take over. Many in the eighteenth century were deists. Today's fundamentalist wrath is not only with evolution but with physics. Without God, is the fear, to be seen with real respect, that our moral structure will fail and with it our civilization. The Inquisition was not foolish when it condemned Galileo for supporting Copernicus' heliocentric views, and who, departing, uttered, "E pur se muove," "And still, it (the earth) moves." The Church rightly knew it faced a huge transformation in the new science that would not only question its authority but might alter the moral basis of Western civilization. But if Copernicus and Kepler moved the heavens from orbit around the Earth with humans at the top of the chain of being, just below angels, then God, at least for Copernicus, there were no *laws* displacing the theistic God. With Newton, all that changed: we are entirely law governed, and the change rings down to the present day.

Art and poetry bounded off the walls of physics and science ever since Newton. An admiration of science gave way with the English Romantics, who recognized that physics with its loss of res cogitans, mind, had cost us our the humanity of our mind, our subjective life. And with its measured objective worldview governed by natural law, it cost us our humanity in that we had become machines, a view hovering with us still today for many thinkers. The Romantics railed against "Science with its rule and line," wrote Keats. Stanley Bates, philosopher, has written: "Why do we still love the Romantics? Because we are still disenchanted!" (Bates, personal communication 2015). We are still disenchanted; Weber was right. In the nineteenth century the role of poetry was to find our soul. By the mid-twentieth century, we found ourselves in a meaningless universe. Consider Steven Weinberg, Nobel physicist, who wrote not long ago, "The more we know of the universe, the more meaningless it appears." His also remained the "Dream of a Final Theory" (Weinberg 1994), the hoped-for triumph of reductive materialism in which all that is or becomes is entailed by natural laws.

If this book is right, reductive materialism fails. Part I will show that no laws govern or entail the becoming of the biosphere, and more, so if Part I is true, reductive materialism fails, for the biosphere is part of the universe, yet is unentailed, so no laws can entail the entire becoming of the universe. In a later, highly speculative Chapter 14, I will doubt even the Pythagorean dream that there *must be* foundational entailing laws. The dream may be true, but we are *not* forced to adopt it, as I will argue, but the argument is very speculative. It suffices to show that we are not forced to the Pythagorean dream; it does not show that dream is false in general but is for the evolution of the biosphere. With reductive materialism, the triumph of Descartes' res extensa via Newton, over his res cogitans, "mind stuff," we have lost our minds and responsible free will. We are becoming either machines or, with classical physics, if we have conscious minds they cannot "act" on the world, so are useless. In Part II, I hope to regain our subjective pole, a responsible free will, and consciousness that can have "acausal consequences" for the world, hence "act acausally on the world," in radical but testable ideas. If Part II is true, we regain our humanity for wide exploration, pursued in Part III where our freedom to co-create what we often cannot prestate yet become into an ever-changing "Adjacent Possible" we unwittingly create, yet must do so with integrity and a reborn spirituality that can be the basis of a transnational mythic structure that may guide us beyond our stubbornly held modernity. We shall need new forms of thinking, a new worldview, new forms of governance, and a widened value system beyond our rabid materialism, a rebirth of our native spirituality, wide concern for what we co-create, value our living "well-discovered lives," lived on our finite planet. Ours is to co-create, but wisely.

But my arguments in this book lie before us to consider, doubt, test, and venture onward.

In the mid-twentieth century, existentialism in a meaningless universe reigned, with Sartre and Camus. We live in such a meaningless universe yet can create our humanity by our choices and actions—a view which, however, requires free will and a mind that can alter the world. Yet reductive materialism denies such responsible free will and mind.

Struggles for and against Newton and his progeny abound in Western philosophic thought. Faced with matter or mind, Bishop Berkeley opted for *mind* in a clear statement of idealism. All is idea. Does a tree falling in a forest make a sound? Yes, it is heard by God. Hegel, responding to stunning Kant who responded to the magnificent sceptic Hume, began an attempt at a Western theory of *becoming*, in place of the Newtonian view of things in being, acted on by external forces. Hegel brings us a theory of history with his famous thesis, antithesis, and synthesis.

In Part I of this book, as hinted, I hope to show that the living world and perhaps parts of the abiotic universe, all beyond entailing law, demand an entire new worldview, and new view of our reality, started by Hegel. As we will see in Part I, living reality consists in ever-new Actual conditions that acausally yield unprestatable ever-new adjacent-possible opportunities that enable, but do not cause, new Actuals to arise, enabling yet new unprestatable Adjacent Possibles in a persistent becoming beyond our knowing. Here we will find History. Moreover, sufficient reason fails us, for we often do not even know what *can* happen. Yet more, we "flow" ineluctably into some of the very possibilities we unwittingly create. Thus, we are free from entailing law, and our co-creativity is, in Part I, our freedom as humans. Wander into a forest and marvel: all came to exist with no entailing laws. How much magic do we wish? William Gaddis said, "There is no truth beyond magic" (1955). We stand to be re-enchanted, as the Romantics so hoped. Beyond Weber, re-enchanted we can surpass modernity.

Modernity is rife with scientism, a certainty that problems can be stated, formulated, and solved. We are mewed up in a dungeon by our mindless scientism. It is largely false, as we will see in Part I. Scientism spawns a cynicism among us as "realists, tough minded, down to earth; just the facts, if you will, problem statement, and solution." All is technologically, or logically, or organizationally statable and solvable. But no, no, no. We are fraught with mystery beyond our design, beyond our intent, and in this, ever at risk, as I show in Parts I, II, and III, are the hope and fear we face as free humans making our largely unprestatable world together.

The Status of the Possible in Classical Physics, Especially in Statistical Mechanics and Probability Theory

I hope to discuss a new view of the world based on a new interpretation of quantum mechanics in Part II, where I will propose that Possibles are ontologically real. This is a pretty wild-sounding idea, but it affords explanations for very strange quantum effects such as "nonlocality" for entangled quantum variables, as I discuss in Part II, and for which we have no current explanations. Thus, it is important to show that even in classical physics, there is an appeal to what seem to be ontologically *real Possibles*.

Hiding in the wings of even classical physics since Newton, as we will see, is our conviction that as classical physicist experimentalists, we *could have chosen to prepare the system with different initial and boundary conditions*. We could have set the canon here, not there. This belief demands that the present "could have been different." In Part II we will see that if classical physics held, we could not so choose. Given the initial and boundary conditions, the present could *not* have been different, so if we are classical systems we are machines and cannot have free will. The present cannot have been different.

Classical physics seems to be a story about what *actually* happens: the billiard balls roll and bounce thus and so. But buried even in classical physics lies a hidden appeal to "the Possible," which shall be central to this book. The boundary conditions, for example, of the billiard table, or a box with gas particles in it, literally defines and thereby *creates* the very phase space of "all *possible* positions and momenta" of the billiard balls on the table, or gas particles in a box, where the box is the boundary condition. If the boundary conditions change, the deduced motions of the balls or particles will change due to the third law, where the balls or particles bounce off the walls and every action has an equal and opposite reaction.

But a strange consequence of boundary conditions is this: in any given set of *actual* billiard balls rolling, say for a short time, on the billiard table, only a few of the points on the boundary will have been hit. Nevertheless, we use the boundary conditions to assert, "Were the initial conditions to "have been different" contrary to fact, the balls "would have hit here and here" contrary to fact, and then the further trajectories of the balls "would have been . . . " In short, we use the boundary conditions to make statements that are "conditionals contrary to fact" that we hold to be true! "Can," "could," and "possible," which are modal statements, enter the entirely deterministic and entailed classical physics. But what is the status of these conditionals contrary to fact?

Notice next that the initial and boundary conditions are essential to classical physics. Without them, we could not integrate Newton's equations. So consider the seven balls on the billiard table, or n gas particles in a one-cubic-liter box. For the seven balls, can "we test" the conditional contrary to the fact that "were the initial conditions different," then the balls would still have obeyed Newton's laws? Yes, we can do so and carry out such tests all the time.

But there is more. We typically cannot create experimentally all possible initial conditions, given the boundary conditions, yet we continue to assert, contrary to fact, that "had" these other initial conditions obtained, Newton's laws "would still have been followed." In short, physicists do in fact think of the set of possible initial conditions and the boundary conditions as literally creating the phase space of "all possible positions and momenta" of the bouncing balls or gas particles. In short, without focusing on it, classical physics is assuming a "reality" to the *possible* positions and momenta of the balls. Even classical physics goes beyond what is Actual and demands a *possible*, the phase space of the system.

The status of the *possible*, central to this book, is even more striking in statistical mechanics, so I must tell you its outlines. Statistical mechanics wishes to "reduce" classical thermodynamics in which "entropy" (useless energy unable to do work) tends to increase. Note "tends to," which is not deterministic, a curious feature of classical thermodynamics and the classical statement of the second law of thermodynamics.

Boltzmann led the way in the invention of statistical mechanics. He sought a theory based on classical nineteenth-century gas particles bouncing like billiard balls off the walls of a liter box. The position of any gas particle can be defined by three numbers, its position on the x, the y, and the z axes of the box. And its momentum, mass times velocity, can be defined by three numbers on the x, y, and z axes. So this gives us six numbers specific to the position and momentum of each particle. If there are many, say n, particles, their current positions and momenta can be defined by $6n$ numbers. Thus, this creates a $6n$ dimensional phase space of all the possible positions and momenta of the n particles in the liter box, which constitutes the boundary conditions. n is very large, on the order of 10 raised to about 23 particles. Boltzmann realized that he could not proceed to integrate the Newtonian equations to follow the actual entailed trajectories of the n particles, where each combination of positions and momenta of the n particles is a single point in the $6n$ dimensional phase space. In fact, given constant energy, the system moves on a $6n - 1$ dimensional "manifold" in the $6n$ dimensional phase space, a point I ignore in this chapter.

The next step is to divide the $6n$ dimensional phase space into tiny, but *finite*-sized, boxes, called microstates, each also $6n$ dimensional. The size of the little boxes is called "coarse graining" and essential to statistical mechanics. But it is the next step that is truly magical. Boltzmann invoked "the ergodic hypothesis" (Boltzmann [1886] 1974). Here one "gives up" integrating the actual Newtonian equations to follow the actual trajectories of the system in the $6n$ dimensional phase space though these microstates. Instead, the ergodic hypothesis simply asserts: "The system will spend equal time in equal volumes of the 6*n* dimensional phase space." But then the ergodic hypothesis entirely gives up "causality." It does not follow the entailed trajectories, a well-known point made to me by Giuseppe Longo. Thus, the ergodic hypothesis is saying that the probability the system will be in any microstate is proportional to its volume. And for that to be true it must first be *possible* for the system to be in that microstate. So, without any causality, the ergodic hypothesis demands an appeal to "The *possible*." One cannot assert "probability" without assuming first "possibility."

Having shown you that the ergodic hypothesis can and *must* appeal to "the possible," I now sketch the second law in statistical mechanics. Define a macrostate, such as all the particles are squeezed into the upper-left corner of the liter box. Count the *number* of coarse-grained microstates in that macrostate. Now consider a more or less uniform distribution of the *n* particles in the liter box. Again count the number of microstates in that "roughly uniform" macrostate. There are vastly more microstates in the nearly uniform macrostate than the macrostate with all the particles in the upper-left corner. So given the ergodic hypothesis, the system will spend more time in the near uniform macrostate than the macrostate with the particles squeezed into the upper-left corner of the box. Therefore, the system *will tend to flow from macrostates with few microstates to macrostates with more microstates.* For simplicity, take the logarithm of the number of microstates per macrostate and call that the "entropy" of that macrostate. Now Boltzmann's results [1886] follow: The system will tend to flow from macrostates of low entropy to macrostates of high entropy.

That is the conceptual cornerstone of statistical mechanics.

I make three points. The ergodic hypothesis is (1) *noncausal* and depends upon an appeal to the (2) *possible,* which has to do with equal probability, or time, spent in equal volumes of the 6*n* phase space of all *possible* positions and momenta, given the box boundary conditions. (3) The ergodic hypothesis is central to statistical mechanics and is not causal, so statistical mechanics does not, in fact, successfully "reduce" classical thermodynamics to efficient cause Newton. But it sure works. Statistical mechanics, in short, demands a reality of the Possible.

The Status of the Possible in Probability Theory in Classical Physics

We all know this. Given a fair coin toss, there are *two possible outcomes,* the probability it will come up heads is 50 percent and tails is 50 percent. (I ignore landing on an edge and remaining there, *a third possibility.*) The same for a fair single dice, the probability of 1, 2, 3, 4, 5, or 6 is 1/6th each. Here there are six *possible outcomes*, ignoring balancing on an edge or vertex of the dice. Physically, with Poincare, let's say that the behavior of the coin or dice is chaotic, so tiny changes in initial conditions can alter the outcome. Then we face "epistemological" ignorance of the outcome. We don't know what will happen, but the behavior in classical physics is entirely deterministic. But watch: for us to be merely "epistemologically" unsure if the outcome will be heads or will be tails, it must be *possible beforehand* that both *possibilities*, heads and tails, are "simultaneously possible outcomes." Unless we say this, it makes no sense to say we are "merely epistemologically uncertain" of which of the two possible outcomes stated beforehand will arise in actuality. Classical physics probability theory demands real Possibles.

In the field of "modal logic" one talks of logically possible. Very roughly, this depends upon the meanings of words. Because a bachelor is, by definition, an unmarried man, it is logically impossible for a bachelor to be married. Another concept is "logically necessary," which is "true in all possible worlds." See David Chalmers (1996), for a discussion of this and the general issue of the nonreducibility of conscious experience to third-person physical statements, the latter a central issue of Part II of this book. In my earlier comments about the "possible" in classical physics, I do *not* mean logically possible worlds. The fair coin really has two possible outcomes, heads or tails (or three with balancing on an edge) stated beforehand.

From boundary conditions creating the very phase space of "all possible positions and momenta" of the billiard balls or particles, to the acausal ergodic hypothesis, to probability theory, it seems classical physics, rather unnoticed, rests on asserting some kind of reality to "possibilities," and even more than one possibility simultaneously. We will see this is spades in superpositions of quantum mechanics on most interpretations of quantum mechanics.

In the next chapter, I will broach an unexpected cousin of the ergodic hypothesis that is a kind of vast "antientropic" process in the universe that I claim has a great deal to do with why the universe became complex, the biosphere along with it as well as our economy and history. This "antientropic process" will be central to a demand for a pattern of explanation for life's becoming and perhaps, beyond Pythagoras, even the becoming of the universe, that is "historical," rather than entailed. My "antientropic process" will not violate the second law as we will see, but the accelerating expansion of the universe via dark energy, that wonderful mystery, probably obviates the worry in any case.

Quantum Mechanics

The first issue to make clear is that I am a doctor, biologist, and philosopher. I am not a physicist, although I have published occasional articles in physics with colleagues and have a founding patent on "the poised realm," a new area of physics hovering reversibly between quantum coherence and "classical" worlds (Kauffman, Vattay, and Nirranen 2014), with Gabor Vattay, a quantum physicist from Eotvos University Budapest and Samuli Niiranen, computer scientist from the Tampere University of Technology Finland and now CEO of a biotech company. The Finnish government gave us a grant of 500,000 Euros in 2012 for work on the poised realm. I will say a lot about physical ideas, mostly concerning quantum mechanics. Many of these ideas are new, all at least partly testable directly or indirectly.

Quantum mechanics will be central to Part II of this book for many reasons. Foremost among these, the causal closure of classical physics leaves no room for "mind" to act causally on the "meat of the brain." Given classical physics, if we have a conscious mind, it can have no effect at all on the world; hence, it is merely "epiphenomenal." Such a mind cannot alter how the tiger chases the gazelle. It is hard to see how evolution would have found such a mind of selective significance. By contrast, quantum mechanics offers two ways in which a quantum mind can have *"acausal"* consequences for the "meat of the brain"; the first is quantum measurement, which, if measurement is real, is almost certainly acausal, and the second is the Poised Realm itself via what is called acausal decoherence and recoherence, as I describe next. A second reason quantum mechanics will be central to us is a recent theorem by mathematicians John Conway and Simon Kochen, the strong free will theorem (Conway and Kochen 2006), which states that if the physicist has free will in a sense I will discuss later, then the electron has free will and "decides" non-randomly what the outcome will be upon its quantum measurement. This is the first theorem I know of which reintroduces "subjective language" with its amazing, "The electron free will decides." The theorem, however, is circular, in asserting that if the physicist has free will, then so does the electron. I will present what I hope are convincing arguments that remove this circularity. If the circularity is removed, electrons do "decide," and we are on the edge of a panpsychism with respect to "deciding" wherever quantum measurement happens in the universe. This does not yet yield "consciousness." With others, I will propose that consciousness is associated with quantum measurement, a hypothesis that is testable in animal models and tentative evidence due to D. Radin (Radin et al. 2012, 2013) that attention at a distance can alter measurement outcomes. But there are further grounds for the hypothesis that consciousness is related to measurement that are related to what is called "the quantum enigma" (Rosenblum and Kuttner 2006), where, to the discomfort of many physicists, quantum mechanics seems to point beyond physics and a purely objective worldview to one which is also subjective. In essence, the enigma states that if we as physicists have free-willed choice about which of two alternative experiments to do, and if conscious observation of the outcome of at least one of these can "induce quantum measurement," as Radin's results tentatively suggest, then our choice of what to measure, thus the question we ask of Nature, and Nature's answer measured by our consciousness, *jointly* co-create reality. This co-creation of reality is truly stunning and is a profound alteration of our view of reality. But I believe I have removed the circularity of the strong free will theorem and we and electrons can "responsibly decide," as we shall see. Furthermore, it is testable whether our conscious observation can "induce measurement." If our conscious observations can sometimes suffice for measurement (but are not always needed, as I shall argue), a solution to the enigma states that electrons in measurement not only decide à la free will but are conscious. If so, the world, the very creation of reality, has a subjective pole including consciousness and free will. I will present a testable potential solution to the quantum enigma. It is highly speculative. It leads to a speculative but possible new view of quantum mechanics, a new triad: actuals, possibles, and mind, conscious and free-willed in us and quantum variables. New Actuals create new Possibles acausally, which are measured acausally by mind to create new Actuals again creating new Possibles for mind to measure, in a persistent vast co-creative becoming of the universe, a status nascent of the universe, biosphere, and economy. Roger Penrose and Stuart Hameroff in their proposed orchestrated objective reduction (Penrose 1989, 1994) of space-time possible superpositions via a gravitational self-interaction are an important cousin to my own proposal, which also leads to a vastly participatory universe as well, as I shall describe in more detail in Part II.

But we shall need something like the classical world for many reasons, including the power of classical general relativity. There are many approaches to the emergence of the classical world, including Stapp (2011) and decoherence (Zurek 1991), and I will have a testable try using the quantum Zeno effect in a later chapter. Given a "classical-enough" world, it can be quantum at base and one hypothesis, mine, is that those quantum variables can measure one another, and do so consciously. Then classical measuring devices can measure, and the moon can be in one place and not smeared out around its orbit when no human is looking.

Quantum mechanics goes beyond seventeenth-century reductive materialism but is still reductive: all that becomes possible is entailed. Nothing novel can become possible.

Quantum mechanics is, with general relativity, one of the twin towers of twentieth-century physics. It has been tested to eleven decimal places, undoubtedly the best-tested scientific theory we have. And quantum mechanics is wondrously mysterious.

Famous physicist, magical genius, and Nobel laureate Richard Feynman states that much of the central mystery of quantum mechanics can be seen in the even more famous "two-slit experiment." Here it is: Consider a barrier, say a thin wooden screen, with two rectangular slits cut in it near one another. In front of the screen is a "photon gun," for example a flashlight, or one with a rheostat allowing the emission of light to be turning up and down. Beyond the screen is an old-fashioned silver halide film emulsion. Cover the left slit with say a small thin layer of wood. Shine the flashlight and later "develop" the film emulsion. You will find a "bright spot" made of many tiny bright spots on the positive image made from the developed film emulsion. On the developed emulsion you will find lots of dark spots. These spots *line up* behind the open right slit, just where you would expect the light to pass through the open right slit and hit the film emulsion.

Now reverse which slit is open; that is, cover the right slit and leave the left slit open and repeat the experiment. You will find a bright spot on the developed film emulsion behind the open left slit, just where you would expect it.

Feynman asks us to consider an analogous experiment with bullets fired from a gun, and two slits, one covered the other not covered, and a "bullet" detector consisting of a sand trap. One would find piles of bullets on the trap behind the open left or right slit.

Now the mystery: Open both slits and shine the flashlight, then develop the film emulsion. You will *not* find two bright spots behind the open left and right slits, but amazingly, you will find vertical bands, light, dark, light, dark, light, dark, extending from where the bright spot behind the open left slit "should be" to where the bright spot behind the open right slit "should be." This alternative light, dark, light, dark, . . . banded pattern is called an "interference pattern."

As Feynman points out with respect to bullets, there is no way classical particles like bullets can yield the interference pattern.

It gets stranger: If we tune down the rheostat so only one photon per hour leaves the photon gun and passes through the two open slits, each leaving a *single* spot on the developed film emulsion, and we accumulate data for several days, then develop the film, we still get the interference pattern. But each photon passes, one at a time, an hour between their flights, on the film emulsion, so however the interference pattern is arising, it is somehow "true" of each single photon.

Quantum mechanics was invented between 1900 and 1927, the most major revolution in our thinking in physics since Newton. It began with the attempt of Max Planck to understand what is called "black body radiation," the radiation yielded by heating a black body. By classical physics, as the temperature of the black body rises, the body should radiate wavelengths that are ever shorter, hence of higher energy. As Einstein argued in his analysis of the photoelectric effect, the energy of a photon is inversely proportional to its wavelength. (Einstein received his Nobel Prize for his 1905 analysis of the photoelectric effect, not special relativity, which was invented during the same year.) But Planck and all physicists realized that as temperature increased toward "infinity," wavelengths of light of ever shorter length, hence higher energy, should, on classical physics, be emitted, leading to an infinite energy at infinite temperature, an "impossible ultraviolet catastrophe."

Classical physics faced the "ultraviolet catastrophe" because it assumed that energy could vary continuously. In a desperate move, in 1900, Max Planck fit an equation to the observed distribution of temperature versus the black body radiation spectrum by assuming to the contrary of classical physics that energy varied in discrete "jumps," which he called "quantized jumps," hence later quantum mechanics.

Meanwhile, Newton had had a corpuscular theory of light, which was displaced soon thereafter by the theory that light propagated as a wave, rather like water waves. Maxwell's field equations in the late nineteenth century formulated electromagnetism as coupled partial differential equations yielding electromagnetic waves propagating, amazing success, with the known speed of light. Maxwell's equations sat poorly with Newton's equations, and in 1905 Einstein formulated special relativity such that light propagates at the same velocity relative to any moving observer in a constant state of motion. In that same "annus mirabilis" in Einstein's life, he formulated his theory of the photoelectric effect, noted earlier. It was known that light shined on certain metals kicked out electrons and did so with higher energy for the ejected electrons if the wavelength of the light was shorter. Einstein proposed that the "wave-like" property of light was somehow also associated with "particle-like" behavior of photons of ever higher energy inversely proportional to the wavelength of the light. The absorption of a photon would transfer a quantum of energy to an electron that kicked the electron out of the metal.

Thus, with Einstein in 1905, we get the first hint of a wave-particle "duality" that remains with us today. Based on these and other ideas, Bohr formulated a theory of the atom in which electrons were in orbits around the nucleus at discrete energy levels. The assumption of discrete energy levels overcame the classical physics issue that classical electrons should spiral into the nucleus radiating energy as they did so. For Bohr, the different energy levels of an electron in orbit around the nucleus accessed by quantum jumps in energy levels, and with the Pauli exclusion principle and Pauling's theory of the chemical bond, became most of the modern theory of chemical bonds for electrons bound by nuclei.

We now come back to the two-slit experiment. Two approaches were taken to this issue: Heisenberg's matrix mechanics in 1926, and a short time later, in 1927, the now famous Schrodinger equation, which remains the foundational equation for quantum mechanics.

The Schrodinger equation is a "linear wave equation" often set equal to a fixed classical potential that is like an infinitely high energy barrier arranged in space in some way. There is no "energy" in the Schrodinger equation, a central point that means that whatever is "waving" in the Schrodinger wave equation is neither energy, or by Einstein's famous $E = mc^2$, matter. It is terribly important that no one knows "what" is waving in the Schrodinger wave equation.

In the Schrodinger wave equations, one or many different waves can propagate simultaneously. Because the wave equation is "linear," the sums and differences of two or more Schrodinger waves can also propagate. This will yield the puzzling "superposition phenomenon" of quantum mechanics.

Of Sea Walls and Gaps

A classical analogy helps understand the application of the Schrodinger equation to the mystery of the interference pattern. Think of a sea wall parallel to a sandy beach. Now consider a train of parallel waves, say 10 meters from crest to crest, propagating from the sea toward the sea wall. Cut two gaps, or slits, in the sea wall. On passing through the gaps, each wave on the beach side of the left and the right gap in the sea wall will propagate a semicircular wave toward the beach. Imagine the first semicircular wave arriving at the beach from the left and the right gaps. Now add the successive waves from the ocean, each sending its own two semicircular waves toward the beach. If you walk along the beach, there will be points on the beach where the crest of a wave from the left gap hits the beach at the same point as the crest of a wave from the right gap. The two waves will add up, or sum, to a higher wave. Similarly there will be other points on the beach where the valleys of two waves from the left and right gaps meet at the same point on the beach and sum to a lower valley. And critically, there will be points on the beach where the crest of the wave from, say, the right gap, meets the valley of the wave from the left gap. The two waves will "cancel," and there is no wave at all.

In this classical analogy, the summed two crests or summed two valleys give rise to the "bright" bands in the interference pattern in the two-slit experiment. The locations where the crest of one wave from one gap cancels the valley of the other wave from the other gap corresponds in analogy to the dark bands in the interference pattern in the two-slit experiment.

Thus, the Schrodinger wave equation has a "variable," called the "Action," which keeps track of the height and location in time and space of each of the possibly many Schrodinger waves. This height and location in time and space of each wave is called the "phase information" concerning that specific Schrodinger wave. In analogy with the sea wall with two gaps, the Schrodinger wave equation spreads through space and time through the two slits in the two-slit experiment. In a very crude but misapplied analogy, just as the crests of two water waves can sum, or the crest and valley of two waves can cancel, the amplitudes of the wave passing through the left slit and that through the right slit can sum to a higher amplitude at some points on the film emulsion. Alternatively, a lower amplitude on the film emulsion can result, or the two amplitudes can cancel, with the crest of one canceling the valley of the other; the Schrodinger wave has no amplitude at that point on the film emulsion.

Here we depart entirely from classical physics. The famous Born Rule, formulated about a year after the 1927 formulation of the Schrodinger equation, states that if one squares the amplitude of the wave (or more precisely the modulus of the wave drawn in the complex plane, where the modulus is the radius in the complex plane of that wave), the squared amplitude is the *probability* that upon quantum measurement, that is the wave that will be "measured." Here quantum measurement *is* the spot appearing at that location on the film emulsion, seen upon development of the film as a dark spot and if a positive image is produced as a light spot.

A deep mystery about quantum mechanics is that there are some fourteen different "interpretations" of quantum mechanics. I will add another one later. General relativity has only one interpretation, so a wonderful mystery is why there are fourteen or more "interpretations" of quantum mechanics. A standard interpretation is the Copenhagen interpretation in which by the Born rule we know *only* the *probability of the outcome of a measurement*, but which wave is measured, as given by its probability, is *ontologically indeterminate*. This ontological indeterminism sets quantum randomness entirely apart from classical randomness. Consider flipping a fair coin. Its dynamics are entirely deterministic but chaotic, so we cannot predict the outcome because we cannot measure initial conditions and, for example, puffs of wind, to infinite accuracy. We are "epistemologically ignorant." But ontologically, the behavior of the coin is entirely determined. In sharp contrast, on almost all interpretations of quantum mechanics, the outcome of a quantum measurement is entirely *ontologically indeterminate*. The randomness is not merely "epistemological" in that, like the chaotic coin whose behavior is deterministic and entailed, we just cannot predict the determined outcome. In quantum mechanics, the outcome is entirely undetermined.

Furthermore, since the same interference pattern occurs when we send one photon per hour through the two slits, the same Schrodinger wave behavior applies to *each* electron separately, and we can know only the probability of the outcome of the measurement of each of the quantum waves, waves without any energy associated with them so we don't know what is propagating.

On the standard Copenhagen interpretation, one says that we have in this behavior a wave-particle duality, first hinted at by Einstein in the photoelectric effect. Before measurement, the quantum system is a "coherent quantum Schrodinger wave" (I will explain coherent in a moment.). *After* measurement there is a stable spot on the silver halide film emulsion seen after development of the film, or on a photodetector array now rather than the film emulsion.

By "quantum coherence" one means that however many Schrodinger waves are propagating, if the squares of all their amplitudes are summed up, they add up to 1.0. So each squared amplitude is a proper probability and their sum is 1.0, as is correct for a probability measure. Hence, the Schrodinger wave is said to propagate "unitarily," meaning the sums of the squares of the amplitudes add up to 1.0 or unity.

The Behavior of the Total Coherent Quantum System Is Entailed

We saw that in Newtonian classical physics, the behavior of the billiard balls is entailed by the initial and boundary conditions and Newton's laws of motion in differential equation form, or by his law of universal gravitation. Newton set our explanatory framework. This is *not* changed by quantum mechanics, for given the initial conditions of the unitarily propagating Schrodinger wave equation, the squares of all amplitudes sum to 1.0, and their time evolution is entirely entailed by the Schrodinger equation. In short, the behavior of the Schrodinger equation is entirely entailed and shows up in the perhaps changing probability distributions of the outcomes of measurement.

This statement is correct until we come to the issue of quantum measurement. There are some interpretations of quantum mechanics, notably the "many worlds" interpretation of Hugh Everett, in which measurement never happens, but at each "measurement event" the universe *splits* into two or more subuniverses that cannot be in contact with one another. However, over the set of all the universes in this multiverse, the total evolution of the Schrodinger wave equation remains entirely entailed, and no new possibilities, that is, squares of the amplitudes of the waves, can ever arise, in this set of branching universes. Within each of the subuniverses, however, no observer can know of the other subuniverses.

I will mention two further interpretations of quantum mechanics. One is due to David Bohm (Bohm and Hiley 1993), who proposed that the behavior of the quantum system is entirely deterministic, by rewriting the Schrodinger equation to yield a new "quantum potential" which does not cause, but guides each quantum "particle" about how to move. On Bohm, measurement does not happen. Instead, particles guided by the quantum potential deterministically arrive at the film emulsion. There is no "probability" of measurement. Bohm's outcomes are identical to all other interpretations of quantum mechanics. In a moment I discuss "nonlocality" in physics and quantum mechanics. Bohm's theory is deterministic but deeply nonlocal.

Finally I stress Richard Feynman's "sum over all possible histories" formulation of quantum mechanics. In brief, in the two-slit experiment, Feynman says that any single photon "simultaneously takes all possible paths through space through the two slits to the film emulsion." I will rely heavily on Feynman later.

For the moment, however, I want to point out a familiar fact of most interpretations of quantum mechanics, including Feynman's. It is seen in the

Schrodinger cat paradox where the cat is prepared in a superposition state. Here a cat is in a box with a poison set to be released if a radioactive random quantum event happens. The paradox is that if we do *not* look in the box, the cat is *simultaneously in a superposition of being both dead and alive.* This is a baby version of Feynman's clear and accepted formulation of quantum mechanics as the sum over all possible histories *before* measurement, in which the photon simultaneously takes all possible paths through the two slits. You see, Feynman is forced to say, with the cat paradox, that a single electron "simultaneously does and does *not* go through the left slit!" But this amazing statement, at the very center of quantum mechanics, breaks Aristotle's law of the excluded middle. For Aristotle, "The cat is on the mat" is true or false. Nothing is in the middle of "true or false," so "The cat *is* and simultaneously is *not* on the mat" is a contradiction. "A and *not* A" is a formal contradiction. The Schrodinger cat paradox and Feynman's formulation of superpositions fail to obey Aristotle's law of the excluded middle.

Fundamentally this failure is exactly why we cannot "picture" quantum mechanical coherent behavior. Furthermore, the failure to obey the law of the excluded middle is why many physicists refuse to assign a "state" to a quantum superposition at all.

The same statements are true both for single quantum variables or for what are called quantum entangled particles in superposition states. In either case the presence of more than one Schrodinger wave implies that both waves are simultaneously somehow more than one possible outcomes of measurement, if measurement is real, or universe splitting on multiple worlds. But superpositions violate Aristotle's law of the excluded middle. The electron can be in a superposition prepared to spin up and simultaneously spin down in the quantum coherent behavior of the system prior to measurement.

I note here and will return later to C. S. Peirce, a famous American philosopher at the turn of the twentieth century, who rightly pointed out that Actuals and Probables do obey Aristotle's law of the excluded middle, but Possibles do *not* obey the law of the excluded middle (Kauffman 2012). Let's try it. "The cat is on the mat" is a proposition about an actual. "The probability of 4,933 heads out of 10,000 coin flips of a fair coin is 0.442" is a proposition about probabilities. We can test it by flipping a fair coin 10,000 times, counting the number of heads, and repeating the 10,000 flips a million times to obtain a Gaussian distribution whose mean is, true or false, near 0.442 within standard error.

So Peirce is right, Actuals and Probables *do* obey the law of the excluded middle. But try, "the cat is simultaneously possibly dead and possibly alive," or "The photon possibly did and simultaneously possibly did not go through the left slit." These statements are *not* contradictions. Next I will baldly and boldly unite Feynman, the mystery of superpositions as in the cat paradox, and Peirce and suggest a new interpretation of quantum mechanics: res potentia, ontologically real Possibles and res extensa, ontologically real Actuals, linked hence united by measurement. This new proposed dualism has testable consequences, as we will see, and is *not* a substance dualism like Descartes' thinking "stuff," res cogitans and res extensa, mechanical stuff. Res potentia are possibles, not "stuffs," so res potentia and res extensa is not a substance dualism, and hence not subjectable to the attacks made against substance dualisms.

Nonlocality in Quantum Physics

In 1935, Einstein, Podolsky, and Rosen sought, in a brilliant argument, to show that quantum mechanics must be incomplete. In quantum mechanics two quantum variables, say two electrons, can be independent or "entangled." If entangled, the two are no longer separate particles, but in a mathematical sense, *one system* described by a single wave function. We need a bit of technical comment. A physicist can use an apparatus to "prepare" two entangled electrons to be measured spin left/spin right *or* to be measured spin up/spin down. Einstein, Rosen, and Podolsky pointed out that quantum mechanics implied that were the physicist to prepare the two entangled electrons to be measured to spin up or spin down for the other, rather than left/right, and then if the two entangled electrons were far apart, say meters or millions of kilometers, the instant one of the two was measured and turned out to be measured spin up, the other would *instantaneously* be spin down. But nothing causal can travel faster than the speed of light, so no causal influence can achieve this instantaneous coordination. "Spooky action at a distance," said Einstein. Bell's inequalities followed some years later. And in the 1990s, A. Aspect in Paris and now very many others have confirmed quantum mechanics. The real universe is *non*local. Correlations arise that cannot be causal. No one knows "how." I will try in Part II to show you that if *possibles are ontologically real*, and an Actual changes, instantaneously and acausally, Possibilities change. If possibles are not located in space, but in time up to special relativity (where is the possibility located in space that you will fall over and hit your head on the floor in 20 minutes?), then we can explain nonlocality and other surprising aspects of quantum mechanics that themselves are rarely noted. Or perhaps Possibles are located in all of space and time up to special relativity. Again, as we will see, we can explain nonlocality with ontologically real Possibles. There is, at present, no account of nonlocality that is accepted.

Bohm's theory, noted earlier, is deterministic by grossly nonlocal. No one has much of a clue about nonlocality. *If* the universe is purely Actuals, and all that happens is *causal* and limited by the speed of light as in special and general relativity, we seem to have no framework to think about nonlocality. Bohm's try is the nonlocal quantum potential, which guides the quantum system but not causally and hence can hope to fit nonlocality, but does not explain nonlocality.

I note that one of the founders of quantum mechanics, Werner Heisenberg, thought that quantum coherent behavior concerned "potentialities" (Heisenberg 1958). So does Zeh, claiming that in quantum coherent behaviors, the "possibilities must be real" (Zeh 2007). I want to hope that my new dualism—res potentia, ontologically real possibilities perhaps outside of space but inside time up to special relativity or perhaps within space and time up to special relativity, and res extensa, Actuals inside of space and time, linked by *real* measurement—provides a possible way to think about nonlocality. I like it, and will discuss it later. But for now, if this new dualism is true, then there can be no mechanism describable by deduction for measurement for simple, but correct, logic: The "x is Possible" of res potentia does not entail the "x is Actual" of res extensa. But this hypothesis is testable. All that has to happen is that someone finds a deductive mechanism for measurement (if measurement is real), and no one has since 1927. Penrose's objective reduction (Penrose 1989, 1994) is just such a proposed mechanism via self-gravitation among superpositions of space-time itself, and it is conceivably testable but does not propose a deductive logic that chooses among the possibilities in the superposition versions of space-time, as I discuss in a later chapter. Indeed, Penrose proposes that measurement outcomes are not random. Nor does Penrose seem to explain nonlocality and other mysteries that ontologically real Possibles seem to explain.

Summary

The aims of this chapter have been to lay out the enormous power unleashed by Newton and classical physics. In this view, we think of the world as entirely governed by physical law, the dream of reductive materialism (materialism because Newton mathematized Aristotle's material "efficient cause" with his forces between particles altering velocities). The conceptual framework is one of initial and boundary conditions and the laws in differential form, whose integration deduces, hence entails, the trajectories of all the particles in the system/universe. The world becomes a machine. If we are such machines, we are machines res extensa. If we are conscious, the causal closure of classical physics precludes our mind from acting on the world. Res cogitans cannot act on res extensa, so our minds are at best epiphenomenal, and it is hard to see why the elaborate brain would have bothered to evolve consciousness that is unable to alter how I get to the market or how the tiger chases the fleeing gazelle.

Quantum mechanics leaves us, ignoring measurement, again with an entirely entailed view of the world in which the initial conditions of the universe yield a unitarily propagating Schrodinger equation whose propagation is entirely entailed. This equation is sometimes called the "wave equation of the universe."

Reductive materialism has been staggeringly successful. We have most of modern technology from it, much of our increased standard of living, wondrous science.

But we have a rampant scientism that plagues us and blinds us. The scientism is part of our "hard-headed realist" stance in a world now devoid of mystery; we are disenchanted and have lost the scientific core of our humanity. We are, in the worst case, mindless robots in a meaningless universe, unable to make even the Existentialists' responsible free-willed choices that give our lives their meanings.

I admire reductive materialism. I seek in the rest of this book not to deny it, but to destroy its hegemony over our minds and set us free. And I do seek to demonstrate that we are not forced to the Pythagorean dream of foundations at all.

PART I

Beyond the Entailing Laws of Physics and Reductive Materialism

3

The Nonergodic Universe Above the Level of Atoms

AN "ANTIENTROPIC PROCESS" IN THE UNIVERSE

IN THE LAST chapter I tried to lay out the conceptual foundations of current physics, starting with Newton, the status of the possible in classical physics, and a beginning foray into quantum mechanics, the focus of Part II in its hope to recover our subjective pole. The central framework of current physics is that of entailing laws. The central image is the billiard table as boundary conditions and the set of all possible initial conditions of position and momenta of the balls on the table. The initial and boundary conditions create, as we saw, the very "phase space" of all possible positions and momenta. Then, given Newton's laws in differential form, we integrate those equations to deduce the deterministic trajectories of the balls. But deduction is entailment, so our framework of science is laws, initial and boundary conditions, and entailment of the consequences. Our model of how to do science, the nomothetic framework of mid-twentieth-century philosophy of science, is to deduce new consequences and test them and accept or reject the results by diverse criteria, then retain or modify our theories. Science proceeds as Aristotle might have wished, in part as deduction. From Hume to Popper to Quine and on, we have debated these issues, specifically induction and deduction as the claimed means to find and test laws.

My aim in this chapter is to begin to show that the Newtonian explanatory framework of initial and boundary conditions, together with deterministic laws of motion in the form of differential equations, which are then integrated to deduce the entailed trajectories of the system, can fail totally even in classical physics and chemistry. My discussion does not rely upon quantum mechanics, although I will mention it. I assume almost exclusively classical physics and chemistry. The point is not to show that Newton's laws do not often work (they do) and better general relativity, nor to go beyond Newton to general relativity, but to begin to demolish the hegemony of reductive materialism and its grip on our scientific minds, and a far wider elicitation of a grossly misplaced scientism in contemporary modernity, whatever we take that modernity to be. Science is *sciencia*, knowing. Being and becoming are more fundamental to all life and our humanity.

Part I of this book wishes to set us free, beyond reductive materialism, surely for the living world, the next two chapters, perhaps for aspects of the abiotic universe, discussed in part in the present chapter, and Chapter 14, which is highly speculative, to help unleash our humanity in a creative universe beyond entailing laws.

Why Is The Universe Complex?

Strangely, at the center of what I want to discuss is a vexed question: since the Big Bang, why has the universe become complex? The universe has become enormously complex: stars, ten to eleventh galaxies each with 10^{11} stars, planets, comets, complex chemistry rising from hydrogen and helium in the first three minutes, to helium, then carbon, oxygen, nitrogen, iron, the most stable atom, up to uranium and a bit beyond, vastly many complex molecules, dust grains, mineral deposits, geologies, and at least on Earth, the evolution of the enormously complex and diverse biosphere.

Why did the universe become complex? Part of the answer is what Paul Davies in *The Goldilocks Enigma* (2006) calls a "put-up job." The laws of physics, general relativity, and the standard model have about twenty-three constants of nature, such as the speed of light, the ratio of proton to electron mass, and so on. Were these constants very different, we could not get a complex universe with stars, galaxies, complex chemistry, and life. This is called the "fine tuning of the constants." An example of this is an unexpected resonance between three helium atoms that allows the formation of carbon and then the more complex atoms in nucleosynthesis.

But there is no known way in physics to derive the constants of nature; we just measure them from outside our theories and put the numbers into our theories.

The "fine-tuning problem" has led to the famous "anthropic principle," which arose when physicists wondered why there are physicists to wonder about why the constants have the values they do. This has led to the idea either that God tuned the constants, or that there are very many universes, with different values of the constants, and we are lucky because we arose, and could have, only in a universe with good values of the constants. In string theory, with its staggering 10^{500} proposed versions, Leonard Susskind has written *The Cosmic Landscape* (2006), where each of 10^{500} universes has a different string theory, and the cosmic landscape concerns how life friendly each is.

In Chapter 14, I will show you that we are not stuck with the anthropic principle if we can conceive of ways the constants of nature could have evolved. I will propose one such way, testable because the constants will have evolved since the Big Bang in a way I hope to describe. Lee Smolin in *Time Reborn* (2013) also considers evolving laws.

The second issue concerns thermodynamics and free energy, that is, the energy available to do work. The second law says free energy is running down. But we know now that the expansion of the universe is accelerating due to the mysterious dark energy that comprises about 70 percent of the energy of the entire universe.

The implication of this accelerating expansion is that we do not have to worry about enough free energy. As the universe become larger, its maximum entropy increases faster than the loss of free energy by the second law, so there is always more than enough free energy to do work.

In short, two parts of why the universe became complex are the fine tuning of the constants and the accelerating expansion of the universe.

But these are only *necessary* conditions for the universe to become complex. They do not tell us why the universe *did* become complex, including the most complex system we know in the universe, our own biosphere, let alone our ever more complex economy.

An Antientropic Process in the Universe

I want to claim that at least part of why the universe has become complex is due to an easy-to-understand, but not well-recognized, "antientropic process" that does not vitiate the second law. Briefly, as more complex things and linked processes are created, and can combine with one another in ever more new ways to make yet more complex amalgams of things and processes, the space of possible things and linked processes becomes vastly larger and the universe has not had time to make all the possibilities. The universe will not make all possible complex molecules, organisms, organs, dust grains, mineral deposits, volcanoes, rivers, geologies, hydrogen clouds, stars, or galaxies, automobiles, or skyscrapers that are possible given the 10^{80} particles in the universe. There is an indefinitely expanding, ever more open space of possibilities ever more sparsely sampled, as the complexity of things and linked processes increases. In a sense that will be central to understanding, for example, biological and economic evolution, that life and the economy evolve into the very possibilities they create as the biosphere or our economy becomes. Often, perhaps typically, these possibilities expand. Every new Actual creates a new set of adjacent-possible pathways or opportunities for further exploration. This holds for space grains evolving in the giant molecular clouds in galaxies, for complex chemistry in space and in the evolving biosphere, for life, for geologies and mineral deposits. Without yet asking if this exploration is deterministic or not, where I will give us grounds both from classical physics and quantum mechanics to doubt that it is deterministic or is so in any way we can know, there is a deep sense in which the universe becomes complex in its exploration of these ever more sparsely sampled spaces of what is next possible because "*it can.*" This "*it can*" is the antientropic process I wish to point out. What becomes Actual enables what next becomes Possible. We will see this explicitly in this chapter in an examination of "hypopopulated, vast chemical reaction graphs with very few atoms or molecules flowing on them." In the next chapter we will see the same thing in the evolution of the biosphere. In the fifth chapter we will see it in the explosion of ways to make a living in the past 50,000 years in our economy. All are expressions of the same antientropic process.

The Nonergodic Universe for Complex Chemical Systems

In this section I will come to classical chemical kinetics as a foundation for what we will explore. But, first, I want to raise a new issue. Consider just CHNOPS, carbon, hydrogen, nitrogen, oxygen, phosphorus, and sulfur, the atoms of organic chemistry. Now consider all possible molecules made of CHNOPS with, say, 100,000 atoms or less per molecule. (Coal is such a molecule and the largest known coal molecule is about $1 \times 2 \times 1.5$ miles somewhere in the United States, a single molecule made only of carbon, with far more than 100,000 carbon atoms.)

The first point is that we do not even know how to count the number of possible molecules of CHNOPS with up to 100,000 atoms per molecules! But it is easy to see that the universe cannot have had enough time to make them all. I give an example I will return to in Chapter 4. Proteins are linear strings of amino acids bound together by peptide bonds. There are twenty types of amino acids in evolved biology. A typical protein is perhaps 300 amino acids long, and some are several thousand amino acids long.

Now, how many possible proteins are there with 200 amino acids? Well, there are 20 choices for each of the 200 positions, so 20^{200} or 10^{260} possible proteins with the length of 200 amino acids. This is a tiny subset of the molecular species of CHNOPS with 100,000 atoms per molecule. Now the universe is 13.7 billion years old and has about 10^{80} particles. The fastest time scale in the universe is the Planck time scale of 10^{-43} seconds. If the universe were doing nothing but using all 10^{80} particles in parallel to make proteins the length of 200 amino acids, each in a single Planck moment, it would take 10^{39} repetitions of the history of the universe to make all the possible proteins the length of 200 amino acids just *once*!

Now consider CHNOPS and all the molecular species with 100,000 CHNOPS atoms. We have no idea how vastly many repetitions of the history of the universe it would take to make them all.

What I have just said is, I think, of the deepest importance. In looking at the ergodic hypothesis in statistical mechanics, which gave up causality, we supposed we could proceed with equal volumes of phase space equally occupied. But as we consider proteins the length of 200 amino acids and all possible CHNOPS molecules with 100,000 atoms or less per molecule, it is obvious that the universe will *never make them all. History* enters when the space of what is possible is vastly larger than what can actually happen. The universe in making a tiny fraction of all possible proteins the length of 200 amino acids is extremely *nonergodic*!

A next point is simple and clear: Consider all the CHNOPS molecules that can be made with 1, with 2, with 3, with 4, with n, with 100,000 atoms per molecule. The space of possible molecules grows rapidly with the number of atoms per molecule. Call the space of possible molecules with n atoms of CHNOPS the phase space for CHNOPS molecules of n atoms. That phase space increases enormously as n increases. Consequently, in the lifetime of the universe, as n increases, that phase space will be sampled ever more sparsely. The universe will make all CHNOPS molecule with 2 atoms of CHOPS, but not all with 100,000.

I will examine later whether the exploration of these phase spaces is deterministic or not, almost surely not in quantum chemistry. But even in classical chemistry, we have no way to study this exploration deterministically, as we will see, even should we wish to do so.

Then there is an antientropic process at work in the chemistry we consider, not one that violates the second law, although given the accelerating expansion of the universe, we may not need to worry about the latter issue.

Classical Chemical Kinetics

I begin our discussion with classical chemical kinetics for a small number of reacting chemical species. In such systems, the typical concentrations of each molecule are on the order of a millimole or micromole, which is a huge number of copies of each kind of molecule, as I show later. I then move on to consider reaction systems with vastly many kinds of atomic and molecular species connected in a vast "chemical reaction graph," where the numbers of copies of each kind of molecule typically range from zero to a few, and show us that we can write no entailing laws of motion in differential equation form for this reaction system which, in principle, cannot explore its reaction space in vastly many repetitions of the lifetime of our universe. Such a system is grossly "nonrepeating" or, again, to use the physicists' phrase, "nonergodic." Because such huge reaction graphs have very few copies of each kind of atomic or molecular species, I will call these systems "hypopopulated reaction graphs." We cannot write deterministic equations of motion for the flow of chemical species on such reaction graphs. Thus, these systems evade "entailing laws" while remaining classical.

A related topic later in the chapter considers chemical reaction systems that are dynamically "chaotic," or at least classically "stochastic," and alter their boundary conditions in ways that we cannot follow over time. The alterations in the boundary conditions mean that we cannot know either the boundary conditions in detail nor how those altered boundary conditions alter the very reaction system for which they are the boundaries. In turn, this will again imply that as boundary conditions change in unknowable ways, we can write no deterministic laws of motion for such a system in differential equation form, and, not knowing the boundary conditions, could not integrate the equations we do not have; so again, no deduced, hence entailing laws for the behavior of such classical chemical reaction systems can be found. We are beyond reductive materialism even in the 350-year-old realms of classical physics and chemistry. We begin to be set free from the hegemony of reductive materialism.

Classical Chemical Kinetic Reaction Systems Among Few Atomic or Molecular Species

It is useful to begin with Avogadro's number: The number of copies of a "mole" of any molecule species is 6.023×10^{23}. Avogadro's number considers the number of atoms or molecules, where the "amount" of each is its "gram molecular weight." Thus, for hydrogen, molecular weight 1, a "mole" is 1 gram of hydrogen. For oxygen, molecular weight 8, a mole is 8 grams of oxygen.

Chemists consider solutions in terms of molar solutions. A molar solution is a mole of a substance dissolved in a liter of a solvent, such as water or ethanol. Chemists typically consider chemical reaction systems which are micromolar to millimolar solutions. One liter of a millimolar solution has 6.023×10^{20} copies of that molecular species. A micromolar liter solution has 6.023×10^{17} copies of that molecular species. A cubic centimeter of solution is $1/1,000$ of a liter and has one-thousandth the number of the relevant atomic molecular species.

The number of copies of each molecular species is so large in the aforementioned cases that, although atoms and molecules are discrete units, chemists can use continuous differential equations for the chemical kinetics of such chemical reaction systems.

The first simple concept is that of chemical equilibrium. Consider two types of molecules, A and B. Let A convert to B at some rate, Ka, proportional to the concentration of A. We can think of this as A -> B at a rate proportional to the concentration of A, so $dA/dt = -KaA$. The "−" sign just means that A disappears in this little differential equation, that is, dA/dt, in each tiny unit of time, at a rate $-KaA$. Let's start with only A present. Soon B is formed from A. But B, once formed, can convert back to A, at a rate, Kb, proportional to its own concentration, yielding $dB/dt = -KbB$. Because we have started with all A and no B, at first B is formed rapidly, but soon starts to convert back to A. At some point the concentrations of A and of B are such that A on average converts to B as fast as B converts back to A. When this happens, the average concentrations of A and B no longer change, and this is chemical equilibrium. Symbolically, $dA/dt = dB/dt = 0$. This equilibrium occurs at some ratio of the concentrations of A and B, and depends upon the ratio of the conversion rates, Ka and Kb.

Chemical equilibrium is our first example of a dynamical "attractor." If a fluctuation occurs away from equilibrium that transiently increases the concentration of B, the rate of B's conversation back to A increases, so its concentration falls toward equilibrium. If the concentration of B fluctuates down, its rate of conversion to A slows while A's conversion to B persists at the higher rate, so B's concentration increases to equilibrium. The equilibrium concentration is an "attractor" of the dynamics of this little chemical reaction system, meaning that trajectories in the two dimension state space of A and B concentrations flow toward the point in the A B state space, which is the equilibrium numbers of A and B molecules. The equilibrium is a stable steady state. Fluctuations away from it due to "noise" return to that equilibrium in a theorem by Einstein, the fluctuation dissipation theorem.

Note again the critical point that the equation for the conversion of A to B and B to A are written as *continuous differential equations*, dA/dt and dB/dt, despite the fact that atoms and molecules are discrete entities in classical physics. These equations can be integrated, as noted earlier, in the two-dimensional A and B state space showing for each point in that space the numbers of A and B molecules, or their concentrations. The integration of these equations yields, from each point, as for the billiard balls on the billiard table, the trajectories of A and B numbers or concentrations over time, leading in the earlier case to a single steady-state attractor, chemical equilibrium. With attention to fluctuations because in fact A and B are not continuous, but discrete molecular variables, we get the fluctuation dissipation theorem. Ignoring the fluctuation dissipation theorem for the moment, the behavior of the chemical kinetic system is entirely deterministic, as is that of the billiard balls ignoring friction. Fluctuations can be added to the chemical systems as classical "noise."

The next concept we need for classical chemical kinetics is the idea of potential wells and potential barriers. Imagine a curved line, with energy on the y axis and "reaction coordinate" on the x axis. High energy is "up" on the y axis. The curve starts reasonably high up near the left margin and dips down to a "potential well" depicting the energy of molecule A at the minimum of this potential well; then the curve rises up over a "hilltop" and dips back down to a minimum, showing the energy of B in its potential well, and then rises beyond this minimum. The basic idea of chemical kinetics concerns the potential hilltop between the wells. For a reaction to occur in which an A molecule converts to a B molecule, the A molecule must "acquire" energy to climb up from the A potential well to the hilltop between the A and B potential wells, the "energy barrier" between A and B molecular species. From this hilltop, the now modified A molecule, call it A*, can fall down to the potential well of B, so converting A to B. The reverse is true: for B to convert to A, it must climb to the same hilltop from its own potential well, and its modified form is B*, which may or may not be identical to A*.

A simple theory says that the probability per unit time at a given temperature of A or of B climbing to the hilltop decreases exponentially with the difference in energy height between the hilltop and the A or the B potential minima. This difference in exponential rates shows up in the rate constants, Ka and Kb earlier. If the energy of the B minimum is lower than that of the A minimum, A converts to B faster than B converts to A, and the chemical equilibrium has more B than A. Note that this theory gives up detailed causality and uses "noise," that is, a "heat bath," and a probability per unit time at a temperature, T, that A or B climbs to the hilltop. The heat bath is classical physics "noise," often treated with stochastic differential equations which draw each "next step" move at *random*, hence "not causally," from some distribution, say a Gaussian distribution of step sizes. Such theories work well, as does the ergodic hypothesis, but give up detailed causality, and hence are not deterministic.

There are very complicated attempts at deterministic theories of the ways A or B can climb to the hilltop which can depend upon the number of pathways up the hill, how these are interwoven, and fluctuations of A or of B to lower energies back down any one of these pathways. Some of these theories show deterministic chaos in which crossing a single hill top is not predictable in details. In short, there are very rich mathematical models of the classical reaction rates between A and B, depending upon the aforementioned factors. Each such model leads to a system of differential equations in the A B state space of numbers or concentrations of A and B, hence to deduced trajectories of the A and B concentrations in that state space. In many but not all cases, these lead to a steady state, as in the equilibrium earlier.

Note that if we consider a one-substrate reaction with two different possible products, A -> B and A -> C, there are three potential wells, one each for A, B, and C, and at least two hill tops, one between well A and B and one between well A and C. Using a heat bath and stochastic differential equation, for any molecule of A, whether it becomes B or C is nondeterministic and acausal, because we have given up integrating the Newtonian equations for the particles in the heat bath, yielded by the sampling of moves from, for example, Gaussian distributions. For chaotic dynamics approaching the hilltops, the results may not be predictable for any single molecule of A. Indeed, chaotic systems cannot be integrated in general. So the formation from A of B or of C is not predictable if we use chaotic dynamics, and not causal or deterministic if we use heat baths.

These tools are in general used to consider reaction networks of tens to hundreds of chemical species, for example in chemical engineering. For a closed system to which no new molecules or energy is added, the system will flow to an equilibrium state, which is a steady-state attractor, subject to fluctuations and the fluctuation dissipation theorem.

In general, the approach to this equilibrium can be by damped oscillations or by trajectories that flow to the steady state without oscillating around it. This is a well-studied subject.

The main conclusion from this is that standard chemical kinetics is studied with continuous differential equations, $dA/dt = \dots$, in the number of copies of a species, here A, or its concentration. Due to the fact that molecules and atoms are discrete, fluctuations occur, dealt with in the vicinity of equilibrium by the fluctuation dissipation theorem that assures mean return to the equilibrium attractor and remaining in the vicinity of this equilibrium attractor in the state space of the concentrations of the N reacting species of atoms or molecules.

Nonequilibrium Chemical Reaction Networks: A Brief Review

Living systems are not closed systems. We eat food and excrete waste. In general, the past decades have seen substantial development in the theory of "nonequilibrium chemical reaction networks," which can have an inflow of matter and energy and an outflow of matter and energy. The topic is vast and not the topic of this book, but I will point at some of the phenomena.

It is useful to classify the main types of attractors of nonlinear dynamical systems, including chemical reaction networks. Such attractors include stable steady states and unstable steady states, which are unstable in one or more dimensions of the state space. Unstable means that if the system is perturbed away from the unstable steady state in an unstable dimension, or manifold, it will flow away from that steady state. A second type of attractor is a "limit cycle," which is a sustained oscillation of the chemical species in their n dimensional state space. The oscillation is a closed "orbit" in state space, hence a "cycle." Trajectories in its vicinity converge onto this cycle and reach it in the limit of infinite time, hence "limit cycle." Wonderful theorems show that in such a system an oscillation often surrounds an unstable steady state. If perturbed from that steady state, the system follows trajectories that spiral outward to the limit cycle. A third type of attractor can be pictured on the surface of a torus. The trajectories can wind around the torus without going through its hole, can wind only through the hole, or can wind both around the torus *and* through the hole. Such trajectories can wind in such a way that after an integer number of windings, *n* through the torus hole and *m* around the torus, the system comes back to the exact same initial state, another limit cycle. But if the number of windings through and around the torus forms an irrational number, that is, is *not* expressible as a rational number whose decimal expansion has repeating digits, then the behavior never comes back to its initial state and is "quasi-periodic." The fourth main type of attractor is "chaotic," which requires three or more variables. Here in the A, B, C, . . . state space, typically trajectories converge onto a chaotic attractor. Once on it, infinitesimally nearby initial points follow trajectories which diverge exponentially in time from one another. This exponential rate of divergence is given by a positive constant, the Lyapunov exponent. Since we cannot measure initial conditions to infinite accuracy, as Poincaré observed for three or more mutually gravitating objects, there is no equation that is an "analytic solution" for the integrated behavior of the system obtained by integration of the differential equations, and "randomness enters," for we cannot predict the long-term behavior. For the attractor to remain in a subregion of its state space despite diverging trajectories, folds occur in the attractor.

The aforementioned systems are spatially homogeneous and described by ordinary differential equations. If we consider space, partial differential equations are used, which is another vast topic. Here, for our immediate purposes, as Alan Turing first showed in 1953, and Prigogine and his group further developed (Prigogine and Nicolis 1977), chemical spatial patterns can arise and be sustained. These theories have been used to try to predict patterns in animal markings, plant and animal morphogenesis, and many other domains. A general term for such systems, coined by Prigogine, is "dissipative structures," of which a whirlpool is an example. Efforts have been made to link the behaviors of such systems to the concept of entropy production, but many doubt the theorems produced. These topics are large and not the main issues I wish to address. Indeed reductive materialism applies very well to such systems.

The Failure of Reductive Materialism in Vast Hypopopulated Classical Chemical Reaction Networks

The standard chemical reaction networks described earlier use continuous ordinary or partial differential equations and can do so because the numbers of copies of each kind of molecular or atomic species is very large, so the concentrations can be treated as continuous variables. But atoms and molecules are discrete entities in classical physics and chemistry. I now turn to chemical reaction systems in which the potential number of chemical species is so large that there is not time enough by many times the lifetime of the universe for all to have been synthesized from any initial distribution of molecules in the system, and in which only a small number, ranging from zero to a few, copies of each type of atom or molecule are typically present at any point in the reaction system. Such systems have not been studied mathematically or experimentally. Thus, I hope to provide a convincing argument that the long-term "actual" behavior of such a system may typically not be describable by any law. By the term "actual" behavior, I mean what "actually" occurred or occurs. We may, in principle, be able to describe what "might possibly" occur, but not always what "actually" occurs."

Chemical reactions can be one substrate one product reactions, two substrate one product reactions, one substrate two product reactions, two substrate two product reactions, and reactions with more than two substrates or products. Each reaction can be thought of as molecules in potential wells with reaction barriers over which the substrates must cross by the accumulation of energy, before the products are formed from the reaction intermediates at the "hilltop." Reactions are more or less reversible, depending largely upon the energy differences between the substrate(s) and product(s) potential well energies. The larger the difference, the less reversible is the reaction.

Chemical Reaction Graphs

We can depict a chemical reaction network "formally" by a special kind of "reaction graph." There are two types of "nodes" in this graph; circles represent atomic and molecular species, and square boxes represent reactions. We will connect substrates of a given reaction by arrows leading from the circles representing those substrates where the arrows are directed into the square box that represents that reaction. Arrows lead from that reaction box to the circles that represent the product(s) of that reaction. I stress that the arrows do *not* point in the direction of the chemical reaction itself; the arrows are only for our convenience and all could point in the opposite direction. The actual flow of the reaction from substrates to products or products to substrates depends upon the displacement, for a closed system, from the equilibrium of that reaction. For chemical systems open to the flow of matter and energy, the direction of each reaction is more complex than mere displacement from the equilibrium of a closed system.

Now consider the famous set of atoms that form organic molecules mentioned before: Carbon, Hydrogen, Nitrogen, Oxygen, Phosphorus, and Sulfur, or CHNOPS. I now ask you again to consider all possible atoms and molecular species made of CHNOPS up to 100,000 atoms per molecule. The limit of 100,000 is arbitrary and can be thought of as far larger.

In fact, as I noted, we have no idea even how to *count* all possible molecules with up to 100,000 CHNOPS atoms per molecule. In part, this is because such molecules can form knots and we cannot count all possible knots. Of course, we do not know this reaction graph. But we can make simple models of this reaction graph. I will then ask us to consider what will happen if we "hypopopulate" it with only a few atoms of CHNOPS, say 1,000, or 1,000,000, or 1,000,000,000. (Remember Avogadro's number is 6.023×10^{23}, so a millimolar solution has about 6×10^{20} copies per liter solution. As noted, micro to millimolar solutions are typical for standard chemical kinetic systems using continuous differential equations.)

I begin with a silly imagined reaction graph to begin to tune our confusion. Suppose that all reactions have two substrates and two products. So two arrows lead into and out of each reaction box for the entire huge reaction graph we imagine. Now let's start our reaction system by placing 1,000,000 atoms or molecules made from CHNOPS so only one copy of each type of molecule is present but only on *one*, not both, of the two substrate circles whose arrows lead to each reaction box.

Will any reactions at all occur? No! No reaction has at least one copy of both substrates or both products, so the reaction *cannot occur*. This simple conclusion is enough to tell us that we must be very circumspect about how matter will "flow" on such hypopopulated vast reaction graphs. The earlier system will not reach equilibrium in any number of lifetimes of the universe, if no molecule can change.

We do not know the actual reaction graph among our up to 100,000 atoms of CHNOPS reaction network. But we can imagine modeling it with some distribution of one substrate one product reactions, one substrate two product reactions whose reverse is two substrate one product reactions, two substrate two product reactions, and higher order reactions. The resulting arrows on the graph, which can be drawn pointing either from substrates to box to products or products to box to substrates, are possible reaction pathways of connected reactions. Let's imagine making a huge ensemble of model reaction graphs with some distribution of the earlier one substrate one product, one substrate two product or two substrate one product, two substrate two product reactions. Although we do not know reality, we can ultimately test which members of this ensemble capture knowable features of real CHNOPS reaction graphs. More the typical or "generic" behavior of matter flowing on each one of such an enormous ensemble of reaction graphs is a new kind of "statistical mechanics" that finds the typical behavior over classes of reaction graphs. Well-chosen ensembles, those that match whatever we now know of real CHNOPS reaction graphs, can, as we refine our ensemble of graphs with increasing knowledge, increasingly show behavior typical of the real CHNOPS reaction graph up to 100,000 CHNOPS atoms per molecule.

We can now study the flow of matter on these model reaction graphs using the famous Gillespie algorithm (Gillespie 1976). Here is how this algorithm works. One begins by listing all reactions, and, for each atom or molecular species, the number of copies of that species. The algorithm picks entirely at random, weighted by the copies of each of the substrates, one of the reactions for which all the substrates are present in one or more copies. Note that by picking *at random* which reaction will occur, we have given up determinism. Then the algorithm picks a "waiting time" for that reaction to occur from an exponentially decaying waiting time that reflects the energy barrier probability of that reaction occurring. Again, by randomly picking from an exponentially decaying waiting time, the stochastic Gillespie algorithm gives up detailed causality. What happens is what is *possible* to happen, drawn from a random choice among the reactions for which there is at least one copy of each needed substrate, and a random choice of waiting time for the reaction. Given the exponential waiting time distribution, reactions tend to happen sooner rather than later. Then the algorithm "makes" that reaction occur, decrements one copy of each of the substrates, and adds one copy to each of the products. Thereafter, the Gillespie algorithm just iterates thousands or millions or billions of times, and the population of molecules in the reaction system "flows" in some way on the reaction graph. The Gillespie algorithm assumes fixed boundary conditions.

Even on simple, let alone huge, chemical hypopopulated reaction graphs, and surely the latter, the Gillespie algorithm (Gillespie 1976) starts from what is Actual, the molecular species present, and picks at random one of the possible next reactions which occurs at a randomly chosen moment and creates a new actual product or products and decrements one copy each of the substrates. This new Actual distribution of molecules sets up the Adjacent Possible next reactions and their probabilities, which depend upon the number of copies of each substrate. I stress that new actuals enable new possibles. This theme will be central to this book.

Now let's think about this algorithm on our enormous hypopopulated reaction graph.

We know virtually nothing at present about the flow of matter on such enormous, hypopopulated reaction graphs. But I want to speculate: For some initial distributions of matter, as in the two-substrate, two-product imagined reaction graph, no reactions at all will occur. For some graphs and initial distributions of matter on the graph, reactions will occur, and old substrates may be decremented from 1 to 0 copies and new products may be incremented from 0 to 1 copy. This explicitly creates new Actuals, the new single copy of the product molecule(s). In turn, the presence of precisely one single copy of one or all of the reactants to a next reaction enables that reaction to happen, that next reaction is now in the Adjacent Possible of the Gillespie algorithm, whose flow is not causal. The process is nondeterministic. Thus, new Actual product molecules can enable new Adjacent Possible reactions which can yield new Actual product molecules which enable new Adjacent Possible reactions. We will find the same pattern the evolution of the biosphere, economy and life, new Actuals enable new Adjacent Possibles into which the system may flow. In Part II, I will come to the same formulation of quantum mechanics with the added testable proposal that conscious observation by us or quantum variables mediates quantum measurement turning Possibles into Actuals.

Furthermore, given the structure of the graph, reactions sequences may flow for a long time and remain in a tiny region of the graph by reversing reactions. For some, probably most, initial distributions of matter on most graphs, the matter will flow widely on the vast graph, *but* if started in exactly the *same* initial conditions, it will visit overlapping or even nonoverlapping regions of the reaction graph as time progresses, sampling the ever-vaster space of possibilities as the complexity of the molecular species increases, but sampling those spaces ever more sparsely. That is, I bet strongly the behavior will be strongly *nonergodic* even from the *same* initial conditions. This presumed variability is *not* standard chaos, from *nearby* initial conditions in a deterministic dynamical system on, say, a chaotic attractor. Here, the very *same* initial conditions on different Gillespie runs, is likely, in the modestly long run, to visit nonoverlapping subsets of the reaction graph. The divergence from the *same* initial conditions is due to the fact that the Gillespie algorithm is not *causal*, in the Newtonian sense of an efficient cause; instead, it picks at random but in a biased way, from the *set* of possible reactions for which there is at least one substrate for each needed input to that reaction, and picks at random when that reaction occurs. Picking a reaction to "occur" at random is *not* causal. Hence, the same initial state can flow in different ways on different runs of the Gillespie algorithm. Note that classical noise, such as temperature, is similarly acausal.

Call each such Gillespie run the *actual* behavior of the reaction system. Can we *deduce*, as if by integration of Newton-like deterministic differential equations, the actual flow of matter on the graph? Again, *no*!

Because the Gillespie algorithm, a rather standard version of a stochastic acausal process, is not deterministic in drawing its next steps from some distribution, the way the system explores the vast reaction graph flowing into sparsely explored areas of the phase space is precisely because *it can*. The new Actuals, that is, the new molecules, do not cause, but enable, the next steps into the new Adjacent Possibles. This is an expression of the antientropic process I am speaking about even in space chemistry in the abiotic universe.

But this conclusion means that reductive materialism fails, if by reductive materialism in classical physics we mean our capacity to deduce what *actually happens*. We cannot. Nor could we *deduce* what *actually happens*, at least typically, were we to know the actual reaction graph among CHNOPS atoms and molecules up to 100,000 atoms per molecule, for a single "flow" of matter on that graph over time. We presumably cannot analytically integrate the equations of motion of such a system even if we could write them down. Surely we cannot integrate those equations if the system is chaotic. Again, classical physics and reductive materialism assert that given the initial and boundary conditions, we *can* deduce what *actually* happens. Here we cannot, either because we use a classical physics heat bath approximation, which gives up detailed causality or any deterministic equations about single substrates crossing two or more potential barriers to form two or more different products if the equations are chaotic. Later I return to chaotic systems that modify their boundary conditions to show that we cannot for long even write the equations for the system, for we do not know how the boundaries are modified.

Is our example physically far-fetched? *No*. Ignoring quantum mechanics for the moment, just these kinds of reaction happen all the time in "space chemistry," including those occurring when meteorites, such as CHNOPS-rich "chondronacious" meteorites, are formed. The Murchison meteorite, which fell in Murchison, Australia in the 1970s, has been analyzed by mass spectrography to have at least 14,000 organic species at the "femtomolar" range, that is, 10^{-15} molar range, which is about 10^8 copies of each kind of molecule. Suppose we could analyze at the "attamolar" range of 10^{-18} moles or about 6×10^5 copies of each molecular species, or down to one copy of each kind of molecular species. What diversity would we find? No one knows but probably it would be very, very large. Space chemistry *is* largely hypopopulated chemistry, for now considered as classical, not quantum, physics, on a vast reaction graph.

Our conclusion, still speculative, are relatively easy to confirm by constructing ensembles of vast reaction graphs, then using the Gillespie algorithm to confirm my earlier speculations of non-ergodic behavior for many initial distributions of CHNOPS, and whether non-ergodic behavior is "generic behavior" for a large class, or ensemble, of still unknown chemical reaction graphs. Such results, if fully confirmed, as I expect will arise, would strongly suggest that we cannot *deduce* what *actually happens* given the specific initial and boundary conditions. Reductive materialism here fails, even in classical physics and chemistry.

Next I will show you that even if we had deterministic chaotic differential equations that modified their boundary conditions, we soon could not write down the equations themselves for the evolution of the system.

Energetic materials may be a second case where classical chemistry cannot deduce what actually happens. If so, reductive materialism again fails.

In the late spring of 2014 I received an email from Leanna Minier at Sandia National Laboratories, Albuquerque, New Mexico. "Dr. Kauffman," she wrote, "We have been unable for over forty years to write equations that allow us to predict the behavior of energetic materials such as explosives and rocket propellants. I want to invite you to be the Keynote Speaker at a Gordon Conference on this topic this summer. We need someone who can think outside the box."

I was puzzled but honored and wrote back that I did not even know where the "box was," but if I learned more, I would be glad to see if I might be of use. Leanna, who heads the group at Sandia studying energetic materials, arranged a day's workshop to orient me, and I later did serve as keynote speaker. I learned enough to begin to understand, but only begin. It may really be the case that we cannot know how to deduce the behaviors of classical physics/chemistry explosives and propellants.

My discussion that follows is in part well-established fact, but at its center it is speculative, awaiting good theorems. I suspect my speculations, presented next, can be turned into such theorems in due course or, in parallel, be studied using the Gillespie algorithm noted earlier.

The first charge to those studying energetic materials is to assess "risk" and safety. For example, such explosives are central to fission bombs, unleashing the compression blast that initiates fission. We no longer test our arsenal, and we would like to know if forty years later they are safe.

Before going further, let's review some statistical information. Some statistical distributions, such as the famous Gaussian or bell-shaped curve, have means and variances. With means and variances, hence "standard deviations," safety and risk can be assessed. But some statistical distributions have neither variances nor means or variances.

Certain "power law" distributions have no variance, or no variance and also no mean. A power law distribution is easily seen in a Log Y vs. Log X plot of the relevant data. If one finds a straight line sloping down to the right as Log X increases, from a high to a low value on the Log Y axis, that straight line says that Y is X raised to a power which is the slope of the straight line down to the right. Mathematically, if the slope is steep, −3 or greater, that is 3 units down the log Y axis for every 1 unit out the log X axis, the distribution has a mean and a variance. But if the slope is flatter than −3, say −2.9, the distribution has a mean but no variance. If the slope is flatter than −1.5, the distribution has neither a mean nor a variance. This mathematical fact underlies Taleb's famous book, *The Black Swan*, where no data tell us what the extreme Log X events will be.

At the conference on energetic materials, I was astonished to find such log plots. These were plots showing, on the Log Y axis, the logarithm of the power impulse applied to an explosive. The Log X axis shows the waiting time for an explosion to occur. The slopes found vary between −0.75 and −0.15, very much flatter than −1.5! In short, for this measure of the behavior of explosives, there is neither a mean nor a variance, so risk and safety cannot be assessed by means we now know.

I explained this in my keynote talk, the point was heard by many present, and these too felt that the facts of these distributions raised serious issues about how to assess risk and safety with respect to any pair of variables showing such a distribution with neither mean nor variance.

The mere fact that a statistical distribution does not have a mean or a variance does not yet show that the underlying causal classical physics mechanism does not allow deduction of what actually happens. But the failure after forty or more years to understand energetic materials is a hint that such a failure may again arise even in classical physics, which asserts that such deduction is possible. But a caveat. In the hypopopulated graphs from earlier, we could study the system from the *same* initial and boundary condition and hope to confirm nonergodic hence no-deducible Actual behaviors. In energetic materials, we cannot assert the same because the actual materials cannot be placed in exactly the same initial and boundary conditions repeatedly. However, I hope next to show that, in principle, an effort to be able to write equations of motion for such systems may well fail and, with it, reductive materialism in classical physics and chemistry. I believe what I shall say is almost surely correct. But caution, near conviction is not proof.

On Chemical Reaction Systems That Create Ever More Unprestatable New Boundary Conditions and Phase Space

Consider first the now familiar billiard table, with its edges the known and fixed boundary conditions that literally *create* the very phase space of all possible positions and momenta of the billiard balls on the table (ignoring balls hopping off the table, or if we wish, put a piece of plywood on top of the table an inch or two above the table surface to keep the balls on the table and under the plywood). If we change the boundary conditions, we *change* the very phase space of the billiard balls; that is, we change the set of all possible positions and momenta of the balls. Following Newton, we must *know* the boundary conditions to allow us to integrate his differential equations of motion for the balls rolling on the table and bouncing off the boundaries. If we do not have the boundary conditions, we do *not* have a mathematical model of the physical system and cannot integrate the differential equations, and so we cannot deduce the trajectories of the balls whose behaviors, not deduced, are therefore not *entailed.*

I now build upon this to construct two classes of systems that change their boundary conditions in ways we cannot follow, so cannot know, so can neither write the boundary conditions to have a full mathematical model we can integrate, nor, worse, since the boundary conditions when altered, alter the laws of motion of the chemical reaction system, we cannot even know the laws of motion of the chemical reaction system when we cannot know how the boundary conditions change. Reductionism will fail again in classical physics if this can be true.

My first example presumes a chaotic chemical reaction system described by continuous differential chemical kinetic equations.

I begin with known boundary conditions; for example, in the black giant molecular clouds in galaxies, one finds tiny and larger dust grains on which surface and near-surface chemistry occurs. The grain is a boundary condition for these reactions, often driven by photon influx. A second example is a small piece of clay surface with chemical reactions occurring on or near the surface. A third example is a living cell, with its bounding bilipid membrane, and a host of internal membranes, structures such as microfilaments, protein pumps buried in the membranes, chromosomes, and a dense maze of protein structures, large carbohydrates, and other macromolecules, and then water, and ionic and other solutes, and a myriad of reactant metabolic and other molecular species. I will focus on the space grain for simplicity.

Now consider the chemical reaction network whose classical kinetic dynamics is, by hypothesis, chaotic. Many such systems are known and require three or more reacting species. By Poincaré's famous result, there *is no analytic solution* to the behavior of such chaotic systems. That is, unlike the elliptical equation Newton found for two gravitating objects, which *is* an analytic solution obtained by integrating his universal law of gravitation for two gravitating objects, there is *no* analytic solution for three or more mutually gravitating objects. More chaos in the sense of "sensitivity to initial conditions" arises, "randomness enters" as Poincaré wrote, because we cannot measure infinitely nearby initial conditions, and the resulting trajectories diverge exponentially in time as given by the famous positive Lyapunov exponent.

The consequence of the fact that there is no *analytic solution* to the integration of the differential equations of motion means that the only way we can study such a system is by numerical simulation on, for example, a large universal Turing machine, or computer. But, critically, any computer has "words" of finite length, representing real numbers of the continuum, to finite accuracy, hence with finite "round-off errors." But these round-off errors mean that as we study the system by numerical integration from some initial condition, the numerically computed trajectory of the system diverges ever more from the true, but unknowable, trajectory of the real chaotic system. This divergence is attested to by the very "sensitivity to initial conditions," which shows that any slight variation induced by a round-off error, given local positive Lyapunov exponents, will send the numerical solution veering away from the true behavior of the system which we cannot know for we have no analytical solution.

In chaotic systems as normally studied, we hold the boundary conditions *fixed*. The behavior of the dynamical system given those fixed boundary conditions is chaotic in the sense mentioned earlier.

But consider now what will happen if, as the system "runs," it alters its own boundary conditions. For example, on the space grain, a given molecule may attach to the grain surface. But if "some period of time" has gone by from the start of our system from its initial conditions, and our numerical solution has diverged from the true behavior of the real system, we *cannot know precisely what molecule bound to the grain surface or precisely where it bound.* That is, we cannot know, due to numerical error, exactly how the boundary condition of the grain's surface has changed. Worse, the chemical species in the reacting system may bind to and come off, the altered feature on the boundary, with on and off "rates." But if we do not know the precise change in the boundary, we cannot know exactly how the on and off rates change, so we cannot write the differential equations for the classical, perhaps still chaotic chemical system. Then we neither know the altered boundary conditions or the new possibly chaotic dynamics of what I will call the "daughter" system of the initial system on the space grain. Now iterate, by hypothesis, the still chaotic daughter system and after some time, and our numerical integration of equations that are no longer precisely those of the initial system nor the same boundary conditions, the now daughter system again modifies its boundaries in, again, an unknowable way. Indeed, we probably know less about how the daughter system modifies its boundaries than the initial mother system, whose boundary conditions and full differential equations yielding known chaos, we knew but could only follow numerically. Thus, over daughter and granddaughter and great-granddaughter systems ever modifying their boundary conditions, we know less and less about the true boundary conditions and the true on and off rates of which chemical in the system that now may bind the new altered boundary. Over time, we can write down neither the altered boundary conditions nor the true differential classical kinetic continuous differential equations for the system.

If this can be true, we cannot deduce the actual behavior of the system from the *same* initial conditions of the mother system of grain and chemical reaction system. Thus, this is *not* familiar chaos, where we cannot deduce the dynamics from the same initial condition because we have no analytic solution to the known differential equations of motion and know boundary conditions. Here we know less and less about the true boundary conditions and true differential equations of motion of the true system as it evolves in time. Thus, unlike standard chaos, we cannot even deduce the behavior of the system from exactly the *same* initial and boundary conditions of the mother system and grain. We do not even know the ever-new boundary conditions nor differential equations of motion, so we cannot integrate them, and can follow their behavior over daughter and granddaughter systems, numerically, with ever decreasing accuracy.

If the aforementioned can be true, classical physics reductive materialism here, again, fails.

But can this be true? The answer must depend in a sense to be worked out, on the spatial scale. Consider a large flat clay surface modified by such a chaotic system in very local ways. If the system is large enough and the modified features small enough, we can "effectively average over" our ignorance and presumably know the ever slightly altered boundary conditions well enough for "effective" differential equations to describe the system "approximately," and stably over generations of slight alterations in boundary conditions. But these are only *effective* differential equations, not detailed causality "entailed" by the equations. Reductive materialism to entailed behaviors fails. Theorems here are badly needed.

Furthermore, it seems likely that on small-enough spatial scales my earlier arguments will hold even more strongly, for the number of boundary features is small, and hence it may well not be possible to "average over" our ignorance and write even "effective equations." Each boundary alteration may change the local dynamics quite strongly, and, if always chaotic, we cannot follow the changes. It is not even necessary that the dynamics be chaotic, only that the changes in the boundary conditions be classically "stochastic" noise that we cannot know in detail.

The aforementioned considerations *may* apply to energetic materials such as explosives. I learned that tiny voids in the energetic materials, whose very size and *shape* matter to the generation of "hot spots" that grow and may grow together to yield an explosion, may be essential to explosion time variability. The tiny voids, whose size distribution is log normal, may have "fractal" boundary shapes, whose fractal dimension in a three-dimensional volume may range from 2.0 to 2.9999. If the voids are tiny enough, it is plausible that their varying shapes may strongly affect the generation and growth of hot spots. For example, needle-like projections radiate heat (and electricity) from their tips ever more strongly as they become more extremely narrow near their tips. Small changes in one or a few of these boundary features in tiny voids may sharply alter explosive behaviors. If so, degrees of freedom on the very small scale, nano and micro, may matter a great deal. I do not know that this is correct for energetic materials, but do note that no one understands why the log input power on the y axis and log time to detonation on the x axis has slopes of -0.75 to -0.15, hence neither mean nor variance.

We don't know, but my suggestions, made at the conference, may be correct. The subject is surely open to study. I tend to think that the idea that we cannot follow altering boundary features and altered chemical kinetic equations in detail on a small-enough spatial scale may very well be correct. If so, we cannot even write "effective equations" and reductive materialism again fails.

I end this section by noting that if the numbers of molecules of each species is small enough, we can use the Gillespie algorithm mentioned earlier, with ever-altering boundary conditions modeled in a still unknown way, to try to study this.

Can We Have a Theory of the Cytoplasm (Cell Biology)?

The cell is a crowded matrix of ordered and disordered water, complex chemical reaction systems, and a myriad matrix of macromolecular components, such as microtubules, microfibriles which move with respect to one another by the action of dynine motors, membranes, and organelles. All of these polymeric structures and surfaces can potentially serve as boundary conditions on chaotic dynamics of the constituent molecular species. By the earlier discussion, if those dynamics alter the boundary conditions in chaotic and thus unknowable ways, over time, we cannot write down either the boundary conditions or the equations of motion. Can we write down the chemical reaction networks and their boundary condition, whether at micro-molar scales or far less? Then can we have a dynamical "theory of the cytoplasm"? The answer is, to me at least, far from clear. We cannot use partial differential equations; the "medium," unlike an iron bar and heat conduction, is not homogeneous. Worse, its ongoing structure and dynamics may be, over time, unknowable. While these are new issues in this chapter and do not yet even address quantum behaviors (see Part II), it is not obvious we have any idea what we are talking about. If we perhaps cannot model in detail explosions, or space grain chemical dynamics, why should we be confident we can have a coherent theory of the cytoplasm, the very nexus of life, classical or quantum or both? If we cannot, what shall we do? We can treat such a failure as merely "epistemological"; we do not know the "real" deterministic, if classical, laws of the cell, but they "exist." Or if we take a pragmatic stance that erases the ontological epistemological distinction, absent in metaphors after all, the "becoming" is just "indeterminate" even in classical, let alone quantum physics.

Conclusions

The purpose of Part I of this book is to end the hegemony of reductive materialism and the hold it has on our view of reality. In that view, in classical physics we are entailed machines, the triumph of Descartes' mechanical worldview over his hoped-for res cogitans, thinking stuff, a substance dualism meant to preserve our subjective pole along with the objective pole of science. The dream of reductive materialism, writ far larger than classical physics, is Weinberg's *Dream of a Final Theory* (1994), whose law or laws, "down there," outside the universe, will entail all that either does actually happen in the becoming of the universe, as in classical physics, or what can happen as in some interpretations, such as the multiple world interpretations, of quantum mechanics where all that can possibly arise is set by the initial conditions of the wave function of the universe. I wish in Part I, and here in this chapter, the first chapter of Part I, to show that this dream is false at least sometimes even in the abiotic universe and fundamentally false in living worlds, the next two chapters. This chapter suffices to cast strong doubt or perhaps is convincing enough in its discussion of the nonergodic behavior of hypopopulated vast chemical reaction graphs to show that we cannot deduce what actually happens. Hence nothing "entails" what actually happens in these cases of classical chemistry. This appears to be an expression of the antiergodic process in the universe I have discussed. I am myself rather persuaded by my argument that at small-enough scales, chaotic systems that modify successively their boundary features and thereby the specific reaction dynamics of the ever-new daughter and granddaughter systems also defy our writing of differential equations of motion, even "effective laws," for such ever-changing systems, so here in the heart of classical physics, reductive materialism can fail. If so, then since we can also use Newton to get to the moon, we must begin to explore when and when we cannot use classical physics as we have dreamed since Laplace as refined by Poincaré and chaotic dynamics. These are new questions that bear fundamentally on the limits of science itself, *sciencia*, "to know." We will see in the next chapter that often we cannot even know what *can* happen in the evolution of the biosphere and, in the following chapter, in the evolution of human life. If so, reason fails us, we are set free from an overwrought scientism that stifles us, and must seek how we live forward in a world when we sometimes or often cannot know what can happen. Even here in Chapter 3, on our hypopopulated chemical reaction graph, we cannot know ab initio from the same initial condition what *will* happen. But we can still know what *can* happen.

In our speculations about chaotic chemical or other systems that change their boundary conditions and features in unknowable ways, we do not even know what *can* happen. *Sciencia* fails, reason fails, and doors will open to how we live forward not knowing what can happen. We start to be set free as humans in a creative universe.

As we will see through Parts I, II, and III, we co-create with one another and nature the very adjacent possibilities, typically not knowing what possibilities we co-create, but into which we become. "There are more things in heaven and earth, Horatio,/ Than are dreamt of in your philosophy."

4

A Creative Universe

NO ENTAILING LAWS, BUT ENABLEMENT
IN THE EVOLUTION OF THE BIOSPHERE

I HOPE TO convince you that reductive materialism stops cold at the magnificent, but beyond entailing law, evolution of the biosphere, the most complex system we know of in the universe. At stake is our very view of reality, surely of we the living and becoming, and perhaps as hinted in Chapter 3 and Chapter 14, speculative, the abiotic universe in many ways. Also at stake is our freedom, and re-visioning that freedom, as alive humans, beyond the grip of Newton, whatever the majesty of classical and even quantum physics, the latter taken up in Part II in seeking our subjective pole. The evolution of the biosphere presents us with an entire new worldview, one in which its becoming cannot be prestated, is not "governed" by entailing laws, in which what becomes constitutes ever-new Actuals that are "enabling constraints" that do not cause, but enable ever-new, typically unprestatable, Adjacent Possible opportunities into which the evolving biosphere becomes. Hegel, in viewing history, saw, roughly, thesis, antithesis, and synthesis. We shall find this theme, one of history, writ large, an expression of the antientropic processes in the universe that I discussed in the last chapter. The most complex system we know in the universe has become in an unprestatable historical unfolding. We are its living children, of this Nature, not above it, not to wrest our due, but to cherish as one sense of sacred.

With this new view of reality, at least at the water edge of evolving life and thereafter, our view of what constitutes how best to do science, one of laws, deduced consequences, then testing of those consequences, must change. This view has no hold on the evolution of the biosphere beyond entailing law. History is richer than entailing laws that do not bind us in all ways.

This work builds upon my own earlier work (Kauffman 2000, 2008), and that with Giuseppe Longo and Mael Montevil (2012) and their separate book, *Perspectives on Organisms* (Longo and Montevil, 2014).

Functions in Biology

In classical physics there are only "happenings." The ball rolls down the hill, bumps a rock, veers, knocking of a chip of rock, and rumbles into a creek with a splash that slightly erodes the bank of the creek. But we say of the human heart: "The function of the heart is to pump blood." Yet the heart causally also makes heart sounds, jiggles water in the pericardial sac, and has myriad other causal consequences. Harvey, before Darwin, worked out that the function of the heart was to pump blood as part of keeping us alive. If we asked Darwin, "Why is the function of Kauffman's heart to pump blood?" Darwin would answer, "Kauffman has a heart because it was selectively advantageous in his ancestors to have hearts that pumped blood and abetted their fitness to have offspring." He might have added that the function of the heart is open to improvement by adaptive evolution via natural selection. Darwin would give a selective account of evolution of the function of the heart. Harvey and Darwin show us that we use "function" all the time in biology. Critically, functions of parts of organisms are typically, if not always, *subsets* of the classical physics causal consequences of a part of an organism.

The concept of "functions" in biology will be central to my discussion. But if there are only causal consequence happenings in physics, in which there is *no way* at all to discriminate a subset of those causal consequences that are "functions," what legitimizes the use of the word "function" in biology? And *if* "functions" are legitimate in biology, we will find that we cannot reduce biology to physics!

I shall answer in two steps, for the concept of "function" is perfectly justified in biology.

First, in the previous chapter I asked us to consider if the universe could make all possible proteins with the length of 200 amino acids, with 20 kinds of amino acids, linked in peptide bonds, in the lifetime of the universe. Recall that with a 200 amino acid length, and 20 kinds of amino acids, there are 20^{200} different protein sequences with the length of 200 amino acids. This number is about 10^{260}. Now, the shortest time scale in the universe is the Planck time scale, 10^{-43} seconds. The universe is about 13.7 billion years old. The known universe has about 10^{80} particles. It is easy to calculate that if all these particles were, in parallel, doing nothing but making proteins with the length of 200 amino acids at each Planck moment, it would take on the order of 10^{39} times the lifetime of the universe to make all possible proteins with the length of 200 amino acids just *once*! Now, has the universe made all possible types of atoms, about 110 or so? Yes. Thus, once we are above the level of atoms to complex chemistry such as proteins with lengths of 200, 300, or 2,000, the universe will never make them all. As I noted in Chapter 3, physicists use the phrase "ergodic" to mean, roughly, that the system will visit all states in a long-enough time. "Ergodic" has a somewhat more technical meaning in statistical mechanics given the ergodic hypothesis that substitutes the acausal ergodic hypothesis that the statistical mechanical system will spend equal time in equal microstate volumes of the statistical mechanics phase space, in place of integrating Newton's equations for the very many particles in the statistical mechanics physical "box" containing the moving particles. In short, in standard statistical mechanics, the meaning of ergodic is that the system will visit all small volumes of state space at some time.

But the universe will not visit all possible protein lengths of 200 amino acids in 10^{39} repetitions of its lifetime. Above the level of atoms, the universe is vastly nonergodic. This has physical meaning; as noted in Chapter 3, history enters when the space of the possible is very much larger than what can occur. The universe will not make all proteins, organelles, cell types, organs, and organisms, in the indefinite hierarchy of complexity of ever-greater nonergodicity up from atoms. The exploration of the ever-vaster space of possibilities as the lengths of peptides and proteins increase, or, as noted in the last chapter, we consider molecules of $n = 1, 2, 3, 1,000, 100,000$ atoms of CHNOPS, becomes ever sparser as complexity increases. This, I believe, is the basis of the antientropic process that undergirds part of why the universe became complex. This is surely true in the evolution of the biosphere from one or a few living things to hundreds of million species today with ever more myriad molecular and interwoven functional diversity of things and linked processes. In a deep sense, the biosphere becomes complex and diverse because *it can*; it becomes into the very possibilities it creates. "Can" here is used either with quantum mechanics or classical physics based either on stochastic equations that acausally draw step sizes from some distribution at random, say a Gaussian distribution. Or the dynamics may be causal but chaotic, so we are epistemologically unable to say what the system "can do." Quantum indeterminate measurement, on most interpretations of quantum mechanics, adds again to "can," as discussed in Chapters 6 through 12 of Part II.

Kantian Wholes and Function

Immanuel Kant, in the *Critique of Judgement* ([1892] 1951), said, "In an organized being, the parts exist for and by means of the whole." He was thinking of living organisms. One way to get to exist in the nonergodic universe above the level of atoms is to be a Kantian whole, in which the *function* of the part is its causal consequences that help sustain the whole. Thus, given a Kantian whole existing in the nonergodic universe above the level of atoms, we *can* define the function of a part as a *subset* of its causal consequences, specifically that or those causal consequences which help maintain the *Kantian whole*.

A specific example of a Kantian whole, also sometimes called an "autopoietic system" that "builds itself" (Maturana and Varela 1980), will help and is related to the issue of the origin of life, which I have worked on from one point of view for decades, proposing in 1971 (Kauffman 1971), that life started as a set of molecules that collectively catalyzed their collective formation from molecular building blocks. Gonen Ashkenasy (Wagner and Ashkenasy 2009) at Ben Gurion University has created such a system of nine peptides (peptides are small proteins), each of which catalyzes the formation of a second copy of the next peptide around a cycle of the nine peptides. This collectively autocatalytic set, CAS, truly reproduces itself in his laboratory. Note the following features: (1) No peptide catalyzes its own formation. (2) The set of nine peptides "collectively" catalyzes its formation from exogenous building blocks consisting of fragments of each of the nine peptides, supplied from "outside." (3) This collectively autocatalytic set is a Kantian whole, or autopoietic system, in which the whole exists for and by means of the parts, where the parts are the nine peptides. (4) We can define the *function* of each peptide as its causal role in catalyzing the formation of the next peptide around the circle. In contrast, if that peptide causally jiggles water in the petri plate holding the nine-peptide system, that causal jiggling is a "side effect" and *not* the function of the peptide. Thus, we have discriminated the function of the peptide, its causal role in catalyzing the formation of the next peptide, from other causal consequences which are side effects. (5) Via evolution of such sets, their functions can sometimes be improved. (6) Critically, this capacity to define a function as a subset of causal consequences that can be improved in evolution further separates biology from physics, which cannot make the distinction among all causal consequences into a subset which are functions. Biology, by this, is beyond physics, and as we will see shortly, because we cannot prestate the evolution of ever-new functions, we can have no entailing laws for the evolution of the biosphere. (7) If we call "catalyzing a reaction" a functional "catalytic task," the nine-peptide system achieves a "functional task *closure*" in its world by which it gets to persist in the nonergodic universe above the level of atoms. "Functional task closure in its world" is a missing, but fundamental concept in biology and, as we shall see, far beyond in the living world. Organisms exist as functional wholes that sustain functional closure, or more generally, a functional "sufficiency" in an effective sense to be defined, in their worlds that may include other organisms to which they are functionally coupled in rich ways. So does the entire evolving biosphere achieve a propagating wave of functional sufficiency along with speciation and extinction events, with, as Darwin said, its tangled bank of interwoven functionalities beyond our current telling. The evolution of the biosphere *is* the unfolding of ever-functional sufficiencies with one another, wave after wave, for 3.7 billion years, with small and large extinction events and now, myriad species and yet more linked structures and process interwoven in that tangled bank.

Think of a bacterium such as *E. coli*. It achieves a myriad of "tasks" or functions such that it manages to reproduce in its world. Among these tasks are DNA replication, construction of membranes, chem-osmotic pumps, organelles, synthesis of proteins, and cell division. All these vast number of tasks are interwoven so the entire cell in its world achieves a task closure, or functional closure, or sufficiency, "in its world." *E. coli* reproduces *as* this functional closure in its world and as a Kantian whole. More richly, consider the 150 bacterial species in the human gut, creating an ecosystem that is critical to human life, and the myriad "tasks" these bacterial species are achieving for one another and us. The definition of "task closure" is hard to make for any single species of bacterium, for its "world" is all the others in that ecosystem, including human cells in the gut. But there is a to-be-defined sense of "sufficient task closure" for each species of bacteria, that the ecosystem of bacteria in your gut and your gut and you get to persist in the nonergodic universe as parts of the ever-evolving biosphere, also able to persist, as it evolves in a persistent, and as we shall see, largely unprestatable becoming beyond entailing law.

Dawkins wrote *The Selfish Gene* (1976) as if DNA replicators were the heart of biology, and the organism merely a "vehicle" for the selfish gene in evolution. I completely disagree. Organisms are autopoietic self-creating wholes that achieve functional sufficiency, often improvable, as the biosphere becomes. Genes, given encoded protein synthesis, play a fundamental *enabling* role in allowing wide exploration of DNA, RNA, and protein space, hence new functions, but it is the functional closure/sufficiencies of organisms as Kantian wholes, not the genes they carry along, which is the heart of life and its evolution. Dawkins was wrong. Life is an ongoing propagation of living interwoven functional sufficiencies over the past 3.7 billion years.

Doings, Biosemiotics, and a Doing, Evaluating, and Action System

Think of *E. coli* swimming up a glucose gradient. This is a causal set that "feeds" the *E. coli*; hence, it is a function of *E. coli* that sustains this Kantian whole in its world. I will define such events as "doings." But the *E. coli* must "sense" its world, and it has done so by evolving receptors for many signals, from glucose to acidity to other features. This sensing of its world's possible states, as given, for example, by the bound and unbound states of receptors for glucose, hydrogen ions, and so on, constitute "biosemiotics" at its root. Once life exists, sensing its world was of selective advantage. But given that sensing, the *E. coli* must "evaluate" "good for me or bad for me," it must make a "decision" to approach food or flee toxin, and then it must be able to act in the world to achieve an instrumental ought. Once doing exists, so do instrumental, not yet ethical, "oughts." This total system is what Peil (2014) calls a self-regulatory system, sensing its world, evaluating it as "good or bad for me," and acting upon that evaluation. Hume discusses the naturalistic fallacy: We cannot deduce "ought" from "is." He is right if one thinks only of *knowing* the world. Yet once "doing" enters the universe, instrumental "ought" is implicit. Given food, approach the food and do so by turning the flagellum using a work cycle and electric motor at the flagellar base. But doing exists throughout the biosphere, so instrumental ought is real in the universe, despite brilliant Hume.

But is *E. coli* just a classical physics evolved molecular machine that can give only the appearance of "choice" and "doing"? This is a deep and further question I take up in Part II. *E. coli* is possibly sentient and has some form of free will. Free will and choice demand that the present "could, counterfactually, have been different." This is not possible in classical physics, bound by what actually happens. But given ontologically indeterminate quantum measurement of superpositions, as I will discuss in Part II, the present could have been different, and free will becomes ontologically possible. Thus, Part II offers a pathway in which we can have consciousness and responsible free will. We humans surely live as if we do. In so far as this is true in evolution, niche construction, well studied, may often involve conscious choices and behaviors. Darwin studied animal emotions. Part II will argue strenuously that we humans are not classical physics machines. Few think our near primate relatives are not conscious, nor lack emotions as part of their behaviors, which have clearly evolved along with morphology. To live with a pet, or watch interspecific friendships, makes it hard to doubt that consciousness and choice may go far down in the animal world or the living world. The possibility of some form of panpsychism, consciousness and free will throughout the universe, is a theme dating at least to Spinoza, and it will be discussed in Part II, with one version proposed by me, and partly testable, another by Sir Roger Penrose and Stuart Hameroff (Penrose 1989, 1994). It is time to open doors, not close them.

We Cannot Prestate All or New Functions That Evolve in the Biosphere

I turn now to the center of my discussion about why entailing laws for the evolution of the biosphere fail. The discussion centers on a new issue: can we prestate all or new uses, hence functions, of "objects" or "processes"? The answer, I now hope to show you, is *no*. If not, we cannot prestate the evolution of new functions in the biosphere, hence cannot prestate the ever-changing phase space of biological evolution which includes precisely the functions of organisms and their myriad parts and processes evolving in their worlds. But these ever-new functions constitute the ever-changing *phase space* of biological evolution. Then if we cannot know ahead of time what new functions will arise, we cannot write differential equations of motion for the evolving biosphere: we have no idea what new entities or processes may arise and become relevant to that evolution. Thus, we cannot integrate the differential equations we cannot write for biological evolution. Thus, we can have *no entailing laws* at all for biological evolution. Furthermore, as I will show, we cannot noncircularly prestate the niche of an organism in its world; hence, we lack both the laws of motion and the boundary conditions, that is, the niche that would allow integration to yield entailing laws. No laws entail evolution.

Consider then, a screwdriver, which I hand to you. Please list for me all the uses of a screwdriver. Well, screw in a screw. Open a can of paint. Wedge a door closed. Wedge a door open. Stab an assailant. Object d'art. Flip it toward the dirt ground in a game where it, knife-like, must land in a circle.

Tie it to a stick and spear a fish. Rent the spear to locals and take 5 percent of the catch. My favorite is: lean the screwdriver against a wall, long edge of end of screwdriver up and perpendicular to wall, lean a piece of plywood on the screwdriver and place a wet oil painting below the plywood to keep rain off the painting. Or scratch your back with it.

I now make two critical claims: (1) the number of uses of a screwdriver is *indefinite*; and (2) the previous list of uses is a "nominal scale"; they are just "different uses." Thus, there is *no ordering* among these uses. By contrast "greater than" is an ordering relation. A thermometer is an interval scale and a ruler a ratio scale. A nominal scale cannot be "ordered" in any way at all.

But if we accept claims (1) and (2), then *no effective procedure or algorithm can list all the uses of a screwdriver or find new uses of screwdrivers*. Further no effective, propositional, procedure can find a new, unordered, use of a screwdriver.

In fact, this is the frame problem in computer science, never solved using propositions, that is, true false statements like the list of earlier uses. The frame problem involves our incapacity to prestate propositionally all *relevant contexts* in which a screwdriver might *find some use*. Artificial intelligence has never solved the frame problem by any algorithm based on propositions, for example a finite prestated propositional list of some uses of a screwdriver, where that list is called a set of "affordances."

Now let's return to *E. coli*, say a single clone of bacteria evolving in some new environment. All that has to happen in evolution is that some molecular or cellular screwdriver in one of the *E. coli* bacteria *finds a use that enhances the fitness of that bacterium in its local world*. Then, if there is heritable variation for that screwdriver, natural selection acting, not at the level of the screwdriver, but at the level of the Kantian whole cell in its world, will probably select a modified *E. coli* with a *new use* or function. Three major comments are critical. (1) We cannot prestate this new use or function, which after selection, becomes part of the evolving biosphere and feeds into the future evolution of the biosphere. (2) This "finding of a new use" *is* "the arrival of the fitter," never solved by Darwin. Darwin could not have solved this issue, for the "arrival of the fitter" is typically unprestatable in its new functionality. (3) While we cannot prestate what selection will select, there are still "winners and losers." Selection does act, but we typically, in the real evolving biosphere, cannot prestate what selection will select, such as new functionalities. Such new functionalities arise all the time, as I show next, with Darwinian preadaptations.

Darwinian Preadaptations

Darwin told us that the function of the heart is to pump blood, not jiggle water in the pericardial sac or make heart sounds, which are mere causal side effects. But he brilliantly noted that such side effects, of no selective use in the current environment, could become of selective significance in some different environment, be selected, and become new functions of organisms. These new functions are called "Darwinian preadaptations," which does not imply foresight on the part of evolution, but merely that these causal side effects "lie to hand" for selection to "seize" and make use of. They are also called exaptations.

I will give several examples of such preadaptations. The flight feathers of birds evolved in reptiles for thermoregulation and were co-opted for flight. The middle ear bones that transmit sound from eardrum to the oval window of the inner ear were preadaptations from the vibration-sensitive jawbones of an early fish. In both cases, new functions, flight and hearing, arose in the biosphere.

I will focus on my favorite example. Some fish have lungs, enabling them to hop from puddle to puddle. Paleontologists believe that water got into some lungs of one or more lung fish. Now there was a sac with water and air in it, poised to evolve into a *swim bladder*, a sac in some fish where the ratio of air to water tunes neutral buoyancy in the water column. Assume the paleontologists are right. This too is a Darwinian preadaptation.

Now with the evolution of the swim bladder, did a new function come to exist in the biosphere? Yes, neutral buoyancy in the water column. Next, did the evolution of the swim bladder *change* the future evolution of the biosphere? Yes, and in two vastly different ways. First, new daughter species of fish with swim bladders and new proteins evolved. But second, *once* the swim bladder exists, it constitutes a new Actual condition in the evolving biosphere. The swim bladder now constitutes a new, empty but Adjacent Possible niche, or opportunity for evolution. For example, a species of worm or bacteria could evolve to live, say exclusively, in the swim bladder. The Adjacent-Possible opportunities for evolution, given the new swim bladder, do not include all possibilities. For example, a *T. rex* or giraffe could not evolve to live in the swim bladder.

The fact that the swim bladder constitutes a new empty adjacent possible niche opportunity for some possible evolutionary steps is of major importance. First, do we think that selection, acting on an evolving population of lung fish, "acted" to "craft" a working swim bladder? Of course, if we will allow ourselves to anthropomorphize "act" and "craft" for the workings of mindless selection. But do we think that selection, in any way at all, "acted" to achieve the swim bladder *as constituting a new adjacent-possible empty niche* in which a worm or bacterial species might evolve to live? No. Selection did not in any sense at all "act" to achieve the swim bladder *as a new* Adjacent-Possible empty niche. But I am staggered by this, for what we have just concluded is that evolution literally creates its own future adjacent-possible opportunities for further evolution without selection in any way "achieving" this. Evolution creates its own future possibilities of the becoming of the biosphere!

We are beyond Darwin, with respect to the unprestatability of the evolution of preadaptations and many other altered functionalities; hence, with respect to the arrival of the fitter, and most important and amazingly, the evolving biosphere literally creates its own possibilities for its future evolution. Life is a miracle of largely unprestatable becoming.

New Actuals arise and do not cause, but *enable* new Adjacent Possibles, new pathways or opportunities for evolution to "explore" in the antientropic process I have discussed in Chapter 3.

This last theme is central in this book. The evolution of the biosphere, of the economy, of legal law systems, of international affairs, of art and poetry and their histories, are all "stories" of new actuals that create new adjacent possibles into which the biosphere and we become. Typically we do not know the very adjacent possibilities we create and then flow into. These histories above the level of atoms where the universe is nonergodic are not just happenings, or willy-nilly; they are largely unprestatable *unfoldings* where what is already Actual creates the possibilities the biosphere and we become.

Further, does the swim bladder, once it has come to exist, *cause* the worm or bacterial species to evolve to live in it? *No*. The swim bladder *enables, but does not cause, the bacterial or worm species to evolve to live in it*. Instead, quantum random mutations to the DNA of the bacterium or worm yield variations in screwdrivers that may be selected at the level of the whole organism by which the worm or bacterial species evolves to adapt to live in the swim bladder. But quantum mutations are ontologically indeterminate and random, and as we noted in Chapter 2 and will discuss in Part II, are due to measurement events that, on almost all interpretations of quantum mechanics, are ontologically *indeterminate*. So the swim bladder enables, but does not cause, the evolution of worms or bacteria, but not *T. rex*, to live in the swim bladder. A new concept, not in physics, has entered: enablement.

Thus, the swim bladder, once it is an Actual that exists, is a "boundary condition," which does not cause, but enables a new adjacent possible in which the worm or bacterium may evolve to live in the swim bladder. (We will see in Part II that the classical potential in the Schrodinger equation is a similar boundary condition that does not cause but enables the behavior of the quantum system.)

No Laws Entail the Evolution of the Biosphere

In physics, we always divide the universe into the system and the rest of the universe. We give the "system" some boundary conditions, like the walls of the billiard table in classical physics that literally define and "create" the very phase space of, here the bouncing billiard balls, all possible positions and momenta of the balls. We thus *prestate* the phase space of the classical physics or quantum mechanical system. This prestatement is essential to our capacity to write laws of motion in differential equation form, and then, also, knowing the boundary conditions, to integrate the differential equations to obtain the deduced trajectories (in classical physics) or quantum waves in quantum. But in the evolution of the biosphere, we cannot prestate the relevant functional variables—flight, swim bladders, hearing—that will become relevant to further evolution. Thus, we can write no differential law of motion for the evolution of the biosphere. We do not know the relevant functional variables beforehand. Furthermore, the organism lives *in its world*. Causal consequences (in classical physics) pass from organism to world and back to the organism, and the functional closure or sufficiency of the organism in its world is what succeeds or fails at the level of that organism in its world. There is, therefore, no noncircular way to define the "niche" of the organism separately from the organism. But that niche is the boundary condition on selection. The "niche" is only revealed *after* the *fact*, by what succeeds in evolution. Thus, we can neither write differential equations for the evolution of the biosphere nor prestate the boundary conditions; thus, we have no laws of motion, and even if we did, we could not integrate them for we lack the boundary conditions. Thus, *no laws entail the evolution of the biosphere.* In this discussion I draw heavily upon the work of G. Longo, M. Montevil, and S. Kauffman (Longo et. al 2012) and on my own earlier work (Kauffman 1995, 2000).

QED

No laws entail the evolution of the most complex system we know in the universe. We are beyond entailing laws, so beyond Newton, Einstein, Schrodinger, and even Darwin.

In a yet larger sense, in reductive materialism, Weinberg dreamt of a final theory (1994), down there which would entail all that does, can, or could become in the becoming universe. But the evolving biosphere is part of that universe. If no laws entail the becoming of the biosphere, then reductive materialism fails in its dream. There can be no such entailing law or laws for the becoming of the entire universe. We can still get to the moon using Newton, general relativity is highly confirmed, and quantum mechanics is confirmed to eleven decimal places. But no laws at all entail the specific evolution of the biosphere. Reductionism and reductive materialism fail as a general view of reality.

With that failure, as we soon see, the mid-twentieth-century model of science, the nomothetic theory, fails in general. In that Newton-like framework, the job of science was to have entailing (mathematical) laws, then deduce new consequences, test those consequences for confirmation or, Popper would say, falsification. Fine for some physics, but not even, if Chapter 3 is right for all of classical physics where we can write no laws. For the evolving biosphere, we cannot even get started on this view of how to do science: we have no laws from which to deduce consequences for the specific evolution of the biosphere and then test them. So too I will suggest in the next chapter, for the evolving economy, systems of legal laws, or history in general.

A New Post-Newton Explanatory Framework

In Newton and quantum mechanics, the boundary conditions literally create the phase space of the system. In classical physics, by Newton's third law, for every action and equal and opposite reaction, the walls of the billiard table play a causal role in the trajectories of the balls on the table. As discussed, noted earlier, in Part II, in quantum mechanics the classical potential in the Schrodinger equation sets the boundary conditions on the quantum system which constrain the behavior of the quantum system which must mathematically "fit" those boundary conditions, but the boundaries play no *causal* role; they create an *"enabled"* phase space.

In the evolution of the biosphere, new Actuals, say the swim bladder, are new Actuals that constitute enabling constraints that enable, but do not cause, new adjacent-possible opportunities for further evolution. But those new opportunities in the nonergodic universe *are* the new potential phase space into which the becoming biosphere literally "becomes." In that restricted set of adjacent possibles, something new that is a new Actual arises, for example, the bacterial species does evolve to live in the swim bladder, creating a new Actual that is again a new enabling constraint that enables a net new Adjacent Possible phase space in which, for example, some amoeba might evolve to poison and feed uniquely or not on the bacterial species now living in the swim bladder. Thus, the new post-Newtonian, post–reductive materialism, postnomothetic model of science (Hemple) is as follows: New Actuals are enabling constraints that create new, typically unprestatable, Adjacent Possibles in which new Actuals may arise, creating yet more new Adjacent Possibles. There is a persistent becoming into typically unprestatable adjacent possibles that evolution itself creates in a radical emergence beyond any entailing laws. Science then also consists in forms of *history*, as familiar already in paleontology: which organisms arose from which organisms? Was there symbiosis? What are the observed patterns of numbers of species per genus, genera per family, up to Phyla per Kingdom? Are organisms on islands like those on the mainlands? What allows speciation? All of these are central biological and evolutionary questions, but they do not partake of deducing from some entailing law the detailed specific evolution of adaptations and in particular new functions in evolution that we deduce ahead of time and test. When we discuss economics in the next chapter, we will find efforts at entailing laws, some of which work modestly well, but we will find that economic innovation is like innovation in the biosphere, typically unprestatable, and creating adjacent possibles into which we become, ineluctably, and typically without foresight about the adjacent possibles *we* enable. The same, we will see, is true of legal systems in which unforeseen loopholes in the law enable new strategies which may call forth new laws and regulations flowing into an unprestatable adjacent possible we enable without knowing fully what we enable when we pass laws or regulations and become into the legal and action spaces we create. The same is true for art history, international relations, politics, venture capital, banking, and life in general. We are radically free and radically emergent and radically cocreative. This, too, is the antientropic process discussed in Chapter 3. Indeed, in human life, we are "sucked into" the very adjacent possibilities we, largely unknowingly, create.

The Failure of Epistemological Closure
in Evolutionary Biology

Some years ago, famous evolutionary biologist Richard Lewontin (Lewontin, personal communication) argued that epistemological closure of evolutionary theory would require (1) a mapping from genotype to phenotype, (2) a mapping from the phenotype via selection to the selected phenotype, (3) and a mapping from the selected phenotype to the selected genotype. But step 2 is not possible. We cannot prestate the mapping from the phenotype to the selected phenotype with a new use for some molecular or cellular screwdriver when selection post fact acting at the level of the whole organism in its world reveals to us what won and what did not win, but we could not prestate what would win and what would not win.

Evolutionary theory cannot be epistemologically closed, a corollary of the lack of entailing laws for the evolution of the biosphere. We are radically emergent, beyond law, and beyond foreknowledge.

If the biosphere, the most complex system we know in the universe, is entailed by no law, it is "foundationless" in the sense of not being entailed by any such law. Then the biosphere does not fit the Pythagorean dream that all that is has foundations, preferably mathematical. In Chapter 14, I will explore the highly speculative idea that perhaps the entire universe is foundationless. It is not impossible, as we shall see. We are not "stuck" with the anthropic principle, or even with the profound conviction that the foundational laws, Weinberg's *Dream of a Final Theory* (1994) are "out there," outside of the universe. Everything may be radically emergent. Maybe.

Reason Fails Us

Once there were lung fish, swim bladders *were* in the adjacent possible of the evolution of the biosphere. Two billion years earlier, with no multicellular organisms, swim bladders were *not* in the adjacent possible of the evolution of the biosphere. So the adjacent possible changes and also does not contain "all possibilities," whatever that phrase might mean. But before swim bladders, we could not prestate the possibilities in the screwdriver enabled adjacent possible evolution of the biosphere given lung fish and the rest of the biosphere. Thus, not only do we not know what *will* happen, we do not even know what *can* happen! If we do not know what *can* happen, reason fails us! We cannot reason about possibilities we do not even know.

Now, consider flipping a fair coin 10,000 times and calculating the probability of 5,340 heads. We use the binomial theorem and calculate the probability, and can test it by flipping 10,000 coins a million times, keeping track of the number of heads each time. We will get a Gaussian distribution centered on some probability which, we hope, is what we calculated with the binomial theorem. But note that while we do not know what *will* happen in our 10,000 coin flips, we do know ahead of time all the possible outcomes of 10,000 flips, all heads, all tails, all 2 to the 10,000 possible combinations of heads and tails. Only because we know this, the sample space, do we have a measure on that probability.

For the evolving biosphere, we do not even know what *can* happen; hence, we do not know the sample space, so we can formulate no probability measure. Thus, we might try Bayesian statistics, but that requires "priors." But priors require something like the set of all the possibles over which we establish our prior probabilities. But we do not know the set of the possibilities in the evolution of the biosphere into its Adjacent Possible. We do not know all the uses of a screwdriver.

We will find the same is true in human life, without or with consciousness and free will, for which I argue strongly in Part II. Then again, reason fails us. It is not sufficient. We cannot reason about that which we do not know can happen. Reason, the highest human virtue of the Greeks and our Enlightenment, is less than sufficient. This raises a profound set of issues: we live forever into a future whose possibilities we cannot fully know. How do we do it? Reason, particularly in the form of true false propositions, fails us some of the time yet live forward we must. How do we do it? What does this mean for our humanity in a creative universe? I return to this in Parts II and III. This is a deep part of our creative humanity in a creative universe.

Jury Rigging and Aristotle's Material Cause

Newton mathematized Aristotle's Efficient cause among his four causes: Formal, Final, Material, and Efficient (Rosen 1991). Material cause is the creation of the house from bricks, from straw, from wood, from stone.

I turn now to "jury rigging" in evolution. Once there is a "function" it can be effectively realized by more than one set of entities and or processes. Consider the Apollo near catastrophe, when a system on board failed. Houston asked the engineers on board to find a solution, and they assembled about twenty objects and went to work and jury-rigged something that solved the functional problem, saving the spacecraft. But what is "jury rigging"? We all know. It is using one or more objects or processes via causal consequences of each alone or together in new ways to achieve a new functional end, not the normal uses of the objects or processes. Once we have functions, many can be roughly jury-rigged in more than one, often myriad ways. Such is the famous tinkering of evolution, the bricolage of which F. Jacob wrote (Jacob 1977). But the capacity to jury rig is just the capacity to build a house from bricks, stones, straw, or boards; it is Aristotle's material cause, banished by Newton and all later science that remains restricted to efficient causes in entailing laws.

But more. Indeed, we will find all Aristotle's causes next. For example, once there are biological "functions," there is a clear sense of teleology, even without consciousness and intent. If we are conscious and have responsible free will, as Part II argues, we choose, full telos. The now exploding field of complexity science seeks what might be called *formal cause* laws of organization that are independent of the specific matter and energy involved. My own theory of the expected emergence of collectively autocatalytic sets is an example (Kauffman 1971, 1986, 1993) and also applies to the real economy, which is a collectively autocatalytic set with input goods going into production functions yielding output goods. The input and output goods are the analogues of substrates and products; the machinery mediating the production function is the analogue of a catalyst, but it is typically produced by the real economy. Formal law theories are independent of the stuff, so are not efficient cause laws. Another example of "formal cause laws" is my own work on random Boolean nets (Kauffman 1969, 1993), which exhibit dynamical order, criticality, and chaos. Cells and the human brain appear to be dynamically critical. The theory described earlier is independent of the specific stuff, the specific matter and energy, whether gene networks or neural networks, so these are not efficient cause laws, but "formal cause laws." To quote Shakespeare: "There are more things in heaven and earth, Horatio,/ Than are dreamt of in your philosophy."

What Is the Ontological Status of the Adjacent Possible in Evolution?

This is a deep question. Were evolution merely classical physics, we might believe, à la Newton and even Poincaré, that all was determined, if not predictable, by entailing deterministic laws in evolution, in a phase space of Possibilities. Chapter 2 argues that even classical physics requires a reality of the *possible* for our use of boundary conditions to make sense, including counterfactual statements, and for the ergodic hypothesis, the actual center to statistical mechanics, to make sense. Classical physics seems to require some sense of *possibles* as *real*. In Part II, I will suggest ontologically real Possibles: res potentia and res extensa linked by measurement, in a new interpretation of quantum mechanics. I will argue that ontologically real Possibles afford a new, perhaps the only or best, explanation for nonlocality and three other mysteries in quantum mechanics.

I have spoken of an Adjacent Posssible. What is its status? This is a complex question.

To restate the issue from Chapter 2 again: standard known boundary conditions in classical physics permit, nay, demand, the truth of counterfactual statements: Not all positions on the boundary are hit by a few balls rolling for a short time, but we say, "Were the balls to have started from different initial conditions, they *would* have hit the walls here and here and Newton's laws *would have described their behavior.*" Thus, boundary conditions, even in classical physics, demand the truth of counterfactuals except for a major issue. *If* the universe were, falsely, classical and started from some initial conditions, say given by God or somehow, and become deterministically, only what actually happened can have come to happen. This holds for general relativity as well, which gives the causal structure of space-time. Therefore, the truth of counterfactuals in classical physics demands different initial conditions, which cannot arise in a standardly deterministic becoming of the universe. But the universe is not merely classical, whatever that is; it is also quantum. Furthermore, in classical and quantum physics, initial conditions are outside of science, as are foundational laws.

Biological evolution is a mixture of "classical physics," quantum mechanics, and selection. On most interpretations of quantum mechanics, as discussed in Part II, quantum measurement is real and ontologically *indeterminate*. Thus, mutations are quantum random and ontologically indeterminate. But then consider a population of fruit flies in the wild with some distribution of genotypes. A mutation happens in one fly, and is selected. Generations later, the sequences of DNA in the later population of fruit flies have changed in some way due in part to unprestatable selection for new uses of screwdrivers. Linked together, this process unfolds into an adjacent possible that is not purely classical deterministic physics, but contains quantum events, almost surely ontologically indeterminate; see Part II. Thus, neither quantum mechanics alone, nor classical physics alone, or, due to the lack of entailing laws, do any laws allow us to deduce the future evolution of the biosphere. I want to conclude that Adjacent Possibles are ontologically real possibilities that arise from this co-mingling of classical, quantum and unprestatable selection in evolution. I argue later from quantum mechanics alone for ontologically real possibles, just mentioned: res potentia and res extensa linked by measurement, a new dualism and interpretation of quantum mechanics in Part II. Thus, there are at least two pointers at an ontologically real "possible." This new dualism is not Descartes' substance dualism res cogitans, thinking "stuff," and res extensa, "mechanical stuff," for Possibilities are not "stuff," as we will see in more detail when I show you that quantum coherent behavior does not obey Aristotle's law of the excluded middle, where obeying that law is necessary for "stuff" and true/false propositions and our worldview prior to quantum mechanics.

Conclusions

We cannot prestate all or the next new use of screwdrivers. Those uses, or functions, stated as propositions, cannot be calculated by any effective procedure or algorithm. But the evolving biosphere finds new functions all the time, in Darwinian preadaptations most notably and in many or most adaptations. Thus, we cannot prestate the evolution of the biosphere in terms of new functions. Those functions are essential features of the ever-changing phase space of evolution. Without prestating the phase space of evolution, we can write no differential equations for the specific evolution of the biosphere, for we do not know the relevant functional variables beforehand, so cannot integrate the laws we do not have, so no laws entail the evolution of the biosphere. Furthermore, we cannot noncircularly specify the niche boundary conditions of an organism, separately from the organism. If we cannot specify the boundary conditions, we cannot integrate the laws of motion we do not have anyway. So again, no laws at all entail the evolution of the biosphere.

If so, reductive materialism, the dream of a final theory (Weinberg 1994) that entails all that does or can happen in the becoming of the universe fails in its largest dreams. The biosphere, the most complex system we know in the universe, becomes radically beyond entailing law in a radical emergence we cannot prestate. Then the hegemony of reductive materialism, despite its enormous and wondrous successes, is over. Over. We are set free at least in the living world, perhaps in the abiotic world as well, as I argued in Chapter 3 and in Parts II and III.

In place of entailing law, we have a new explanatory framework in which new Actuals arise and are enabling constraints that do not cause, but enable, ever-new, typically unprestatable Adjacent Possibles, whose possibilities we cannot know beforehand. The evolving biosphere *becomes* into these new

Adjacent Possibles, creating yet new Actuals creating yet new Adjacent Possibles, and evolution, without selection "acting to do so," itself creates the very Adjacent Possibles into which it flows. All is becoming, all is radically emergent, all is beyond entailing law. We will find the same in much of human life. Sufficient reason is not sufficient. Then how do we live forward, not knowing what can happen?

This unprestatable becoming is also an expression of the antientropic processes of the universe I discussed in Chapter 3.

We are set free from constraint to entailing law, which renders us machines, even before we consider the mind-body problem in Part II. We, like the biosphere, co-create the very possibilities, often with little or no foresight with respect to the adjacent possibilities we create then become. Reason is insufficient for life, more mysterious and wonderful than we have supposed in our scientific, scientistic modernity. In the next chapter I take us to our human life, yet do so before defending consciousness, mind, free will, and these subjective poles of our humanity in a creative universe in Parts II and III.

5

A *Creative Universe*

THE SOCIAL WORLD

THE HUMAN LIVING world is open, radically emergent, governed by no entailing laws, and, like the biosphere, flows into an often unprestatable and largely unintended Adjacent Possible whose shape we often do not foresee, but co-create. Organizations, like organisms, have functional closures or sufficiencies in their worlds and adapt in often unprestatable ways to live or die. These themes apply to economic life, legal life, political life, international relations, the history of art, and history itself.

Yet despite this, we believe to a perhaps astonishing extent that our world is knowable, optimizable, and controllable. This conviction drives overconfidence in top-down control, reliance on "experts," and how we govern ourselves as our organizations adapt unprestatably into the Adjacent Possibles they and we create, often for their own interests, forming our power structures that throttle us. We need far more bottom-up enablement wisely coordinated, but hardly know how to do so. What is it we would wish to co-create to the extent we have choice and live our evolved and cultural lives together?

Economic Theory

Economic theory assumes sometimes responsible free will, rationality, consciousness, and preferences. I will defend these in Part II, and here assume them.

Our often overly mathematical economic theories of ourselves are also often deeply inadequate.

I begin by sketching three major themes of modern standard economic theory, "infinitely rational economic man," "competitive general equilibrium," and "game theory." All are foundational to standard economics. All are inadequate.

Much of contemporary economic theory needs to assume that economic agents are "infinitely rational." This assumption allows economic agents to "solve" intensely hard optimization problems that economists propose are parts of real economic life. Of course, "infinitely rational" is just silly, and Herbert Simon long ago proposed "satisficing" rather than optimizing. The difficulty is that, in well-formulated optimization problems, we can formulate mathematically what an infinitely rational agent might do to optimize his or her expected utility, based on his preferences, while there are indefinitely many ways to be less than infinitely rational. In the past decades the growth of behavioral economics has shown that we are not, of course, infinitely rational, and this work is finding its way into economics. But infinitely rational is needed for the central theorem of modern economics, competitive general equilibrium, by Arrow and Debreu (1954), discussed next.

Competitive General Equilibrium

A foundational question in economics is how prices come to be "set," such that, by and large, supply and demand are jointly satisfied and "markets clear," that is, the goods and services produced are typically also purchased at some list, or "vector" of prices, one per good. Here, markets "clear," which constitutes the economic notion of "equilibrium." It is an amazing fact that most major cities have three days' supply of food and are confident that the "market" will, by and large, keep the markets filled. That economic markets largely "clear" is a real and major issue.

Most of us are familiar with supply-demand curves. As the supply of corn goes up, the price per bushel falls, and as price falls, demand for corn goes up, so the curves of supply and demand in a Cartesian coordinate system cross at some point where supply equals demand, and the market for corn "clears." The concept of "equilibrium" in economics, again, is just "markets clear."

Walras, a Swiss economist in the nineteenth century, likened supply-demand curves to a potential function in physics, such as a bowl, where a ball in the bowl rolls to the bottom of the bowl. A hope that economics could be made to resemble physics has grown in much of the economics profession, whose papers and articles are typically wonders of mathematical sophistication. As we will see, however, we cannot even mathematize much of economic life and the evolution of the economy.

The concept of supply-demand curves becomes more complex once there are two goods, say bread and butter. Changes in the supply and demand for bread may alter the supply and demand for butter, for many of us want to put butter on our bread. At what prices for both bread and butter do the markets for these two goods clear? Then how do we think about an economy with very many goods that are to be traded. Is there a list, or vector, of prices such that the supply and demand for all goods are met and markets clear?

The answer can be yes, as formulated in competitive general equilibrium, a brilliant theorem by Arrow and Debreu (1954), which won the Nobel Prize, and remains at the foundations of modern economics. It takes a bit of thinking to get one's mind around this theorem.

Here are the central ideas and results. We assume from the outset that we can *prestate* all possible "dated contingent goods." A "dated contingent good" is, for example, a bushel of wheat delivered to your door next July 14, *if* it rained in Nebraska on July 2 of the same year. So we are to imagine all possible, but prestated, dated contingent goods. But some of these dated contingent goods are out in the future, for example next July 14. So the second idea is that all economic agents involved are infinitely rational and will contract for these goods, each having his or her own probability distribution for each of these dated contingent goods, whose distribution may be very complex and vary over time for each good. The next assumption is that there is a way to trade each and all of these dated contingent goods, that is, that there are markets for *all* dated contingent goods. In the theorem itself, the behavior in real time of such markets is sidestepped by assuming an auctioneer who, at the "beginning of time" and "once and for all time," auctions off bids and asks among a set of economic agents in front of the auctioneer, for each of the possible dated contingent goods. The brilliant theorem then demonstrates that there does, in fact, exist a list, or "vector" of prices such that contracts are purchased at those prices for each of the dated contingent goods and, given the probability distributions over future contingencies held by all agents, *however* the future really unfolds, all the contracts whose contingencies are met are fulfilled and all markets "clear." This wonderful theorem *is* competitive general equilibrium (CGE). It answers Walras's dream.

CGE is a foundation of economic theory for it proposes that there is a "yes" answer to the allocation of scarce resources in which all markets clear. Within the economics profession, this theorem has been subject to enormous examination, criticism, and extension, much beyond the scope of this book. But I sketch a few well-known issues: (1) Do all markets for all dated contingent goods really exist? If not, what happens? (2) Incomplete knowledge, where different agents have different knowledge. (3) "Sunspot" equilibria, where there is more than one vector of prices where markets clear. What chooses among these equilibria, where some equilibria make most of us better off than other equilibria?

There is a fundamental reason for these concerns. Economists would have liked to believe that there was a single equilibrium, and that market forces sufficed to find this optimal solution, which was optimal in diverse senses for the general community. This is a version of Adam Smith's "invisible hand" metaphor. Along with many other issues, the hope that market forces would lead to a social optimum, defined as optimizing the "utility function" of each economic agent, for example, fails. Nothing guarantees a "social optimum."

Competitive General Equilibrium
Is Almost a Nonstarter

I raise a major, then an even more major, point. First, could we in fact, prestate all contingencies such that we could prestate all dated contingent goods? We find new uses of screwdrivers all the time. We cannot prestate those uses and hence their possible utility to economic agents; hence, we cannot prestate the contingencies that may arise in the future that we do not know beforehand can arise. We cannot prestate "all dated contingent goods." Even for a fixed set of goods and production functions, CGE is rather a nonstarter. It cannot be completed, for we cannot list all dated contingent goods.

The second issue is related to the first but more radical. We cannot prestate the new goods, services, and production capacities that will be invented. Economic innovation is largely unprestatable and emergent. This is a major theme of this chapter, and I return to it later. The unprestatability of economic innovation implies, as does the unprestatable emergence of new functions in the evolving biosphere, that we cannot prestate the phase space of economic life, that is, all dated contingent goods and production capacities, so we can write no entailing laws for the evolution of the economy. Again, to be examined later, like the biosphere, new goods and production capacities are new Actuals that are enabling constraints that do not cause, but enable new, typically unprestatable Adjacent Possible economic opportunities into which the economy "flows," creating again yet new unprestatable new Actuals that enable yet new Adjacent Possibles. Economic evolution is open-ended, creative, beyond entailing law, and radically emergent.

Game Theory

A second major strand in economic theory is game theory, invented by von Neumann and Morgenstern. It is beautiful, but totally inadequate to the real world.

The central ideas of game theory are these: We assume some game, such as the famous "prisoner's dilemma of "cooperate" or "defect." Two prisoners are held in two rooms by the police, accused of a crime. There is a prestated "payoff matrix" showing for each of the two prisoners what happens, his or her payoff, if she cooperates or defects *and* if the other prisoner cooperates or defects. "Cooperate" means not "ratting" on the other, that is, confirming the other's role in the crime. Defect means "ratting" on the other. In the present game, here is the payoff matrix. If both prisoners "cooperate," that is, do *not* tell the police that they "did the crime," or do not confess, they both do *not* go to jail. If either one, but *not* both, of the prisoners defect, that is, confess that they both did the crime, while the other prisoner cooperates, that is, denies the crime, then the defecting prisoner gets two years in jail, and the cooperating prisoner gets ten years in jail. If they both "defect," that is, confess to the crime they jointly did, they both get eight years in jail. Note that "both defect" is eight years in jail for each, while if one defects and the other cooperates, the defector gets only two years in jail, and the cooperator gets ten years in jail, which is worse than eight years in jail if both defect. The "dilemma" is that both are better off if *both* cooperate, and they both do not go to jail. But neither can know for sure that the other will not defect, so faced with this uncertainty, each prisoner is better off defecting, assuming the other will cooperate. She will only get two years in jail while the other, cooperating, gets ten years in jail. But both face the same dilemma and both may defect, so both get eight years in jail.

Mathematician John Nash formulated the concept of solutions to games, called pure- and mixed-strategy Nash equilibria. In a pure-strategy Nash equilibrium, both players are best off, each consistently playing one of her possible "strategies" or "actions," here "cooperate" or "defect." In a mixed-strategy Nash equilibrium, both players are best off playing some combination of their strategies or actions, with some probability distribution, effected randomly over repeats of the same game.

Game theory has led to much work, including generalization to *n* person games with multiple strategies, and been generalized to evolutionary biology by John Maynard Smith in the concept of "evolutionary stable strategies," which are strategies among two or more organisms or species that cannot be invaded due to selective advantage by mutants that alter the stable strategies. These are lovely concepts.

But game theory is inadequate to the real world in its supposition that one can *prestate* the payoff matrix of the game along with a prestated set of actions or strategies. A lovely example will begin this part of my tale. The Peruvian government, seeking at last to protect the Amazon forest, passed a law stating that the forest canopy would be monitored by satellite, and that any person disturbing the canopy would be punished. In response, the locals began to cut down trees *shorter* than the canopy and sell the timber! What had happened? In response to the new law, the locals invented a new "action," "cut down trees shorter than the canopy." We might or might not have been able to prestate this new action. But such "new actions" arise all the time. For example, the NASDAQ stock market has many rules for market makers from whom the public buys and sells stocks. Almost inevitably, some market maker finds a way to circumvent some rule by an unforeseen new "action" or strategy. The NASDAQ "police" may step in to control the trader evading the rule, but also other market makers may use the same new strategy to circumvent the rules for their own trading advantage. Ultimately, new rules are devised and the cycle repeats.

Thus, I claim that we cannot prestate the actions that a law or rule or regulations may *enable, via unprestatable loopholes*, as new actions or incentives to actions (Devins et al. 2016). Caryn Devins (Devins, personal communication 2014), while a law student at Duke, wrote her senior thesis on the unintended consequences of minimal sentencing for powder and crack cocaine. Many black communities, fearing the harm crack might do to their communities, favored such minimal sentencing. Thinking that crack was ten times as dangerous as powder cocaine, the law initially stipulated a tenfold longer sentencing for crack than for powder cocaine. And, critically, to make the law tough, the law asserted a minimal sentence. There were manifold unintended and unanticipated consequences. White kids in private bedrooms snorting cocaine could not be searched without warrants for probable cause. But black youngsters selling crack on the streets to make money in impoverished areas with few job opportunities were easy marks for the police, who won "good marks" for arrests. Thus, this law changed the incentive structure for the police. So lots of black kids were arrested. Then the tough minimum sentence for crack created a powerful further incentive for prosecutors to get easy convictions, and brownie points professionally, by *plea bargaining* with blacks accused of selling crack. While racism may have much to do with the consequences, our jails are now overfilled with young black men on drug charges.

Devins rightly argues that the law creating stiff minimal sentences for crack did create a new, unforeseen "action strategy" by prosecutors: emphasis on plea bargaining, not trial by jury, to obtain convictions. More generally, good lawyers facing complex laws are masters at finding *unexpected loopholes* in those laws. Then such unprestatable new loopholes open unprestatable new opportunities and thus open up, hence *enable*, but do not *cause*, new and typically unprestatable actions, strategies, or behaviors. The payoffs of these new strategies are themselves typically not prestatable. In short, laws both constrain some actions yet enable other, often unforeseen actions with unforeseeable payoffs.

With respect to game theory, these examples of the use and abuse of legal laws already tell us that we cannot prestate either the set of actions or strategies in a game, nor can we prestate the payoff for each of the players. Thus, in real life, games evolve into new games in ways we cannot prestate, with new unforeseen loopholes in the case of legal laws and regulations, enabling new, often unforeseen actions, behaviors, or strategies, with often unforeseeable payoffs. In general, then we face an open-ended evolution of games, here laws, loopholes, Adjacent Possible opportunities and thus Adjacent Possible strategies, in real life.

Laws and strategies are a generalization of social norms and roles, stable or evolving, so our social forms can and often do evolve as well in unprestatable ways. Anthropology studies norms and roles. The history in any culture of the evolution of these, and where the alterations were or were not prestatable, and from what "sample space" of possible strategies and actions, seems very much worth consideration. I doubt we can prestate the sample space of emergent strategies and payoffs.

Organizations are quite like organisms, functional wholes seeking to persist in their worlds, and adapting in often unprestatable ways to the opportunities that present themselves in their worlds, thereby creating new Actuals that enable, but do not cause, new and typically unprestatable Adjacent Possible opportunities into which these organizations separately and together co-evolve, ever creating new adjacent possibles. We are almost inevitably drawn into the ever-changing Adjacent Possibles we co-create.

Of course, organizations are people living and working together. As individuals, we have our own personal interests, negative and positive emotions that undergird many of our reasons, purposes, motives, and decisions, knowingly or not. But we also adopt the aims of our organizations. The chief coach for the Yankees dearly wants his or her team to win and is fulfilled by such winning. The CEO of a bank too big to fail acts, as recent history demonstrates, for the interests of his bank, blatantly selling "short" for the bank's interests, the very same assets the same bank sells "long" to its customers.

The adaptive evolution of organizations into the Adjacent Possibles they and we co-create, typically unknowingly has, I think, ever more profound meaning as the rate of formation of the Adjacent Possible, the ever new opportunities, blossoms. We are "sucked into" the very opportunities, possibilities, we co-create. Sometimes this is good, sometimes it leads to the metastasis of power and greed in our adapting power structures co-opting regulations and their loophole interpretations for their own interests, thereby creating yet further Adjacent Possible opportunities for good or ill.

The Creative Economy

In the evolving economy, new goods and services, and new production capacities, sometimes including new unforeseen uses of old goods and production capacities, create new Actuals that constitute boundary conditions that enable, but do not cause, new adjacent-possible economic activities, goods, services and production functions flowing into an unprestatable and ever-changing Adjacent Possible.

In 1933, Alan Turing invented the famous Turing machine, the first clean mathematization of "mechanical computation." It fostered amazing theorems, such as the undecidability of the "halting problem," a parallel to Goedel's famous theorem on formal undecidability in rich-enough axiomatic systems. Soon thereafter, von Neumann invented the von Neumann computer architecture at the Institute for Advanced Study. The first computer, the Eniac, was created at the University of Pennsylvania during World War II to calculate shell trajectories for the US Navy. The invention of the universal computer enabled, but did not cause, IBM to make and sell universal mainframe computers, where its CEO expected sale of no more than three. But the wide sale and use of mainframe computers created a market for the personal computer, which was invented soon after the chip was invented. That market enabled, but did not cause, the widespread sale of personal computers for myriad uses. Personal computers were Actuals that enabled an adjacent possible that included word processing, and Microsoft and others entered that market. Wide use of word processing enabled file sharing. Widespread use of file sharing became an Actual that enabled, but did not cause, the invention of the World Wide Web. The new Actual, the Web, enabled a new Adjacent Possible, that included selling goods on the Web, and eBay and Amazon came into existence. These and other uses of the Web were new Actuals that put content on the Web which enabled, but did not cause, a new Adjacent Possible that included Web browsers, and Google and Yahoo came into existence. So did Facebook and other social media on the Web, from which Facebook played a role in the Arab Spring.

A few years ago a Japanese man living in Tokyo in a tiny apartment with many books and a new baby realized he could scan his books into his new iPad, sell the books, and have more space in his apartment for his young family. *Then* he realized his new business opportunity. Tokyo is filled with tiny apartments, given its crowding. Many lived in such tiny apartments. The Japanese man realized he could go to these people, scan *their* books into *their* iPads, sell their books, and take 5 percent of the sales as his profit. His new business flourished and is now copied. What had he done? He had found a "new use" for the iPad: scan other people's books into their own iPads, sell the books, and take a profit. Do we think Apple thought that the iPad might create this new business opportunity? No.

Recently, Google has invented Google glasses with chips that can read online, can take pictures, and many other things. Google is "crowdsourcing" its glasses, sending them out to many computer scientists and others to find *new uses* for Google glasses! Business is beginning to make use of the very unprestatability of the adjacent possibles it knows it creates.

Notice that for the Japanese man, and Google, both could not prestate their market niche, rather like our incapacity to noncircularly prestate the niche of an organism separately from the organism as a whole making a living in its world. Venture capitalists hire good management teams because they know that the initial business plan will likely fail as the new business seeks its market, that is, it seeks its "niche" that it often may not know beforehand, and a good management team abetted by more venture capitalists and other funding may live to find it.

Could competitive general equilibrium with its prestated set of all possible dated contingent goods even gain a foothold on this emergence into an ever-new, usually unprestatable Adjacent Possible? No. CGE is a nonstarter with respect to radically emergent economic innovation.

Could we have entailing laws for the open-ended, emergent, creative evolution of the economy and technology? *No.* We do not know beforehand the variables, that is, goods, services, and production capacities, that will arise. We can write no laws of motion in differential equation form for this evolution any more than we can for the evolution of the biosphere.

Economic evolution is an unfolding history in the nonergodic universe above the level of atoms, a further expression of the antientropic processes

I discussed in Chapter 3. Because the adjacent possible has some, but not all possibilities, we are enabled, post fact, to reconstruct that history and make sense of what happened, what was enabled at each step. New Actuals afford new Adjacent Possible opportunities which, once taken, yield new Actuals. Economists look down on economic history, preferring theory and theorem. Well, fine, theory and theorems are to be desired where possible. But we make a mistake if we think we can mathematize the detailed specific evolution of the economy. Our First-World economy is, by and large, exploding into an exploding Adjacent Possible.

Autocatalytic Economic Webs

Fifty thousand years ago, the diversity of goods and production functions in the global economy was perhaps a few thousand. Now in New York one can find over a billion goods and services. The diversity of the economy has exploded.

Why? Think of an economy as a graph with two kinds of nodes, circles and squares, just like the chemical reaction graphs of Chapter 3. Here the circles are goods and services, and the boxes are production capacities. For each box, or production capacity, there are typically one or more input goods. Two boards and a nail are input goods to a production capacity that is a hand and a hammer. The person with the hand, using the hammer, nails the two boards together, creating a product, two boards nailed together. Depict this with arrows from the two input goods to the production function box, and an arrow from the box to the product good, the two nailed boards.

Now notice that the hammer is like a catalyst, a chemical that speeds a chemical reaction, and the hammer is used to nail the boards together but is not "used up" in nailing the boards together. But the hammer itself is a product of the entire economy. So the economic web of goods and services is in a generalized sense, collectively autocatalytic, and makes use of input goods such as iron ore, mined from the ground.

Earlier, in Chapter 4, I noted that Gonen Ashkenasy at Ben Gurion University has a nine-peptide collectively autocatalytic set that reproduces itself (Wagner and Ashkenasy 2009). In a later chapter on the origin of life, I will discuss the now well-worked-out theory of the spontaneous formation of collectively autocatalytic sets as one theory for life's origin (Horkijk, Steel, and Kauffman 2012). It is remarkable that the same set of ideas, collective autocatalysis, applies to chemical and economic systems. The "general theory of collectively autocatalytic sets" concerns objects, transformations among objects, and objects helping or hindering those transformations. The theory is fully independent of the "stuff" constituting the objects, or the processes mediating their transformations. This kind of theory is entirely unlike Newton's efficient cause entailing laws that do depend specifically on the stuff and forces among the stuffs. I believe we can call such laws "formal cause laws," echoing Aristotle's formal causes. In fact, "complexity theory" has grown enormously over the past several decades and was sometimes derided as "fact-free science." Why? Because complexity theory sought laws of organization, such as collectively autocatalytic sets or dynamical criticality (Kauffman 1993), that are independent of the specific stuff, matter and energy. We can learn about the world in ways far beyond reductive materialism. This is a very big issue. Newton taught us to use efficient cause entailing laws, and the nomothetic theory of science (Hemple) saw only this as how to do science. But this view is just wrong. The economy is, in part, a collectively autocatalytic set. So are cells.

The economic web has exploded over the past 50,000 years. I asked, Why? In part this is due to "recombination." Old goods and services can sometimes be broken apart and recombined in new ways. A parachute attached to the tail of a Cessna becomes an airbrake. Then the richer the economic web already is, the easier it is to find new combinations. The Wright Brothers' first airplane is, in part, a recombination between an airfoil, light gas engine, propellers, and bicycle wheels. But more than recombination is occurring. The Japanese man with the iPad found a *new use*, screwdriver-like new use, for the iPad and founded a new business offering a new service: scan your books into your iPad and I will sell your books and take 5 percent. Thus, the economic web also can expand in a way that cannot be prestated. In part, we can mathematize the prestatable aspects of this explosion; in part, we cannot mathematize its detailed becoming into an ever-expanding adjacent possible as the web itself expands.

The Economic Web and Economic Growth Theory

Economic growth theory largely ignores the economic web and the role that its own diversity plays in its own creation of and growth into the very adjacent possible it creates. Growth theory is usually a theory of a single-sector economy making "stuff" such as GDP, with inputs such as labor, capital, human knowledge, and some ideas about knowledge "spillover," savings and investment. Fine work has been done, but it pays no attention to the fact the web itself creates new economic opportunities all the time, as the Japanese man and Google glasses being "crowdsourced" to find new uses for their glasses demonstrate.

This is a big issue. We have sought to foster economic growth based on the Washington consensus: achieve stable government, money supply, law, and infrastructure, and growth will follow. Sometimes growth does follow. Much of the world is in poverty. In part, understanding better the roles of the economic web might help us help poverty. In my book, *At Home in the Universe* (Kauffman 1995), I discuss a simple model of subcritical and supercritical economic growth. Plot on an x axis the diversity of goods and on the y axis the diversity of ways these can be (algorithmically only so far) combined to create new goods. A hyperbolic curve in this space separates subcritical behavior below that line and supercritical behavior above that line. Below it, few new goods are created; above it an explosion of new goods can come into existence if economically profitable. Add to this algorithmic story one with unprestatable new uses and goods and production functions, and we must think of this as a new part of economic growth theory.

Beyond Economics

The unprestateability of what we enable when we act and, for example, pass laws and regulations has its parallels in international relations, where we often do not know what we will co-create. The European powers stumbled into World War I. No one knew how World War II would flow. No one foresaw the collapse of the Soviet Union, nor how the post–Cold War era would unfold. We all adapt for our own and to some extent our joint interests in ways we can only partially prestate. History unfolds into the possibilities it creates, new Actuals enabling new Possibles enabling new Actuals.

A current example is "terrorism." While I fear the overuse of this ill-defined term for power issues, we do indeed face terrorism. Terrorist organizations and responding governments are adaptive organizations and will adapt in unforeseen ways. The military has the concept of the OODA loop, basically the decision loop of assessing the situation and acting upon it. A central problem, as I have learned from high experts in the US military and government, is that terrorists react and adapt so rapidly that they are "inside" or faster, than our own "OODA" loop. We play catch-up to their adaptive moves.

We think we can "design institutions," but even our US Constitution has been strongly deformed in the arguments between "original intent of the Founders" and the "Constitution as a living document." Caryn Devins, having clerked for the Chief Justice of the Vermont Supreme Court, recently gave a talk against design (Devins et al. 2015), pointing out how the Commerce Clause of the Constitution regulating Interstate commerce has been widely used, and perhaps co-opted, for the War on Drugs. Institutions grow, adapt, and morph for myriad reasons. No one designed English common law. It evolved by precedent. A wonderful book by Stewart Brand, *How Buildings Learn* examines how English manor homes evolved over generations as owners found new uses and opportunities to grow their homes in unexpected ways. We become differently than we intend.

Larry Arnhart, in *The Evolution of Darwinian Liberalism* (2014), discusses two of the three strands that led to our current mythic structure (beyond Newton and entailing laws), Adam Smith, and Darwin. In place of the Church's view before the Renaissance of a God-given order from outside upon which civilization and social structure could be built, Adam Smith gave us "the invisible hand" where the shoemaker and pin maker, acting for his or her "selfish" benefit, would lead to the benefit of all. Social life could be self-organized from within. Darwin gave us not only evolution by natural selection but an initial oversimplified image, of selfish evolutionary optimization, later seen in social Darwinism, or survival of the fittest, or, not Darwin's phase, "Nature red in tooth and claw." It is an image of selfishness, akin to Smith, from which a natural order emerges from within. Of course, these images overstate "selfishness" and ignore cooperation, empathy, caring, and collaboration. But as centerpieces of our mythic structure, Arnhart makes a good point. These themes underlie much of modern Western First-World social life and political ideals.

But "self-organization from within" is a sprawling, interwoven, wondrous, emergent mess in the evolving biosphere, within and among organisms and the abiotic world, and in our social, organizational, and political lives as individuals, institutions, rules, norms, roles, strategies all evolving, partly statable and predictable, partly not. We must live into the future and make it, structured by our past, yet not fully knowing what we enable when we enable. Philip Clayton, philosopher and friend, commented to me: "How shall we garden the Adjacent Possible?" "How and for what civilizational ends?" (Clayton, personal communication). I would add, for this freedom and confusion is just the point of this book, that we are living humans, creative, in a creative universe.

PART II

Toward the Subjective Pole

6

Are We Zombies with, at Best, Witnessing Minds?

IN PART I, we came to a view of ever-new Actuals enabling ever-new Possibles enabling ever-new Actuals in a continuous becoming, surely for the evolving biosphere, and economy and human life, and probably, in Chapter 3, for aspects of the abiotic universe above the level of atoms where as complexity increases, the phase space of possibilities becomes ever vaster and is explored ever more sparsely. This process is enormously nonergodic and is the antientropic process I discussed in Chapter 3, part of why the universe may have become complex. All this is a becoming, a status nascendi. I will come in Part II, over the next chapters, to much the same view in my attempt—be careful, I am not a physicist—at a new interpretation of quantum mechanics. I will propose a new Triad: Actuals, Possibles, and Mind, to be carefully defined herein, where mind *acausally* observes Possibles mediating quantum measurement, transforming Possibles to new Actuals, which then *acausally* enable new Possibles for mind to observe again, hence in a continuous status nascendi. This Triad, I will propose, includes quantum variables, such as electrons exchanging photons, or perhaps fermions exchanging bosons, consciously observing and measuring one another and acting with free will, and human conscious, free-willed mind. This will lead to a radical panpsychism; wherever measurement occurs, so do consciousness and free will. The concept that quantum variables can measure one another, whether consciously or not, is central to this proposal, but is not the standard view of measurement. This hypothesis may be testable, perhaps by what is known as the quantum Zeno effect, as I describe later.

I emphasize that the Triad proposal makes new predictions beyond standard quantum mechanics. In particular, the proposal that mind acausally "mediates" measurement is in principle testable. One way would be to show that human conscious attention *non*locally, hence acausally, can alter the outcome of measurement. I explore nonlocality further later. As I also describe, Radin et al. (2012, 2013) has very tentative evidence that humans distant from a two-slit experiment, while paying attention, but not otherwise, can alter the ratio of light and dark interference band intensities. Each spot in the interference pattern is a measurement, so altering the intensities *is* altering measurement. Were this confirmed, and were it ever shown to occur nonlocally, mind must acausally, due to nonlocality, mediate measurement, and must be in some sense quantum to account for non-locality.

I am not alone in proposing a panpsychism. Sir Roger Penrose and Stuart Hameroff have proposed a somewhat similar view (Penrose 1989, 1994) explored more fully later in this chapter in which measurement is associated with flashes of consciousness and choice. On these views, we are not zombies at all, but have consciousness and responsible free will. On this view, consciousness and free will are part of the fabric of the becoming of the universe, which is thus emergent and creative because of free will choices from quantum variables upward. If this is testable, and it is in several ways for my views and those of Penrose (1989, 1994), and if confirmed, we will have to consider that consciousness and free will did not emerge *with life*, but as part of the universe, like pressure and temperature, was used as life evolved, possibly sentient and acting from the start, and that consciousness became ever more integrated and refined and diversified, as we ourselves experience. Some of what I discuss next is presaged in overlapping ways in Kauffman (2011, 2015).

We will also have to discuss approaches to the "classical" world, given quantum mechanics. Stapp (2011) has one approach; decoherence (Zurek 1991) is another. I will propose a third, based on the quantum Zeno effect later. It seems testable. Evolution could not have happened without a stable-enough classical world for adaptations to accumulate. Nor could we explain the vast non-random diversity of Linnean-related fossils in the geological record without a classical-enough world for evolution to have occurred. Nor could we have "embodiment" of living organisms.

But many have come to think we may be zombies. Why?

Descartes set the stage for modern science in the period around 1640 with his famous substance dualism: res cogitans and res extensa. Res cogitans was "thinking stuff," and res extensa was the external third-person objective view of a mechanical world. Given his *Discourse on Methods* (Descartes 2014), Descartes could hardly avoid taking consciousness seriously, for his famous

Method was to doubt everything. However, he found he could not doubt that he was doubting: "I think, therefore I am" was his starting uncontestable proposition. "I think" is for Descartes conscious thought. Thus, his entire effort rested on consciousness.

But skeptics emerged rapidly: how could res cogitans, "thought," affect the real world, res extensa? Descartes proposed the pineal gland as the locus where res cogitans altered the brain, body, and world. The idea did not work then and we live with its consequences today.

In this chapter I would like to review briefly our struggles since Descartes and the triumph of Newton. We have lost the subjective pole, central to our humanity, while gaining the objective pole, and have not yet found our subjective pole in science as we know it. Part II is an effort, some testable, to find a framework that might suffice. It will lead us in very radical directions to a universe in which consciousness and responsible free will and doing and a real physical world are all parts of the entire universe down to the particle level at least, thus constituting one form of panpsychism. Do not accuse me of being a Mysterian. I know most of the issues and will be clear. Indeed, as noted, Sir Roger Penrose and Stuart Hameroff have proposed a kind of cousin (1989, 1994) to what I will write, and I will discuss their work briefly later, where flashed of consciousness and free will emerge at quantum measurement. My Triad, Chapter 12, will propose Actuals, Possibles, and Mind (conscious and willing) mediating measurement from fermions exchanging bosons to us, converting Possibles to Actuals. Penrose and Hameroff and I separately propose a vastly participatory universe that is conscious at measurement and choosing, so doing. This will be supported by the strong free will theorem in Chapter 11.

The famous physicist John Archibald Wheeler speculated about the universe observing itself (Wheeler 1990). If the universe merely observed and "witnessed" itself, with no further consequences of such witnessing, why would the universe "bother" to merely witness itself? Of course, "bother" supposes the universe itself has purposes, and who knows, perhaps it does, but I mean much less, merely that a universe that just observed itself is less rich than one in which that observing is coupled with "choosing" and "doing," that alters what happens in the becoming of the universe. But observing and doing is, for us humans, just what we suppose for our own subjective pole, where at least sometimes what we consciously observe is linked to what we sometimes with responsible free will *do*, that therefore alters the world. We built a rocket, but think we could have chosen otherwise, and chose to send it to Mars, but could have done otherwise, altering the mass of Mars, so the dynamics of the solar system. What I will say approaches Wheeler. Be deeply skeptical about much in Part II. Much is clean, some is testable as we would want in science, and the ideas hang together in large part and answer several mysteries about quantum mechanics, including nonlocality. As we will see, it is hard to find a way forward with our standard ideas. But caveat lector.

The Stalemate: The Causal Closure of Classical Physics

Reconsider with me the Newtonian worldview. As we have discussed, Newton formulated the differential and integral calculus, three laws of motion, and universal gravitation in a fixed space and uniformly flowing time. Consider again seven billiard balls rolling on the billiard table, whose boundaries constitute the boundary conditions and act causally on the moving billiard balls via his third law, "For every action an equal and opposite reaction." Then were we to ask Newton what would happen with our seven rolling balls, he would say, roughly, "Please write down at any instant of time the positions and momenta of all the balls, the initial conditions. Please write down the boundary conditions. Now my three laws, in differential equation form, give the forces between the balls as they hit one another, via $F = ma$. To find out what will happen to the balls over time, please *integrate* my differential equations given the initial and boundary conditions and the results of that integration will be the trajectories of the balls on the table for all time, ignoring for now, friction." (Later friction was included.) But we can now note that to integrate the differential equations of motion is to *deduce* the consequences of those differential equations of motion for the trajectories of the balls on the table. But deduction *is entailment*, as is "All men are mortal, Socrates is a man, therefore Socrates is a mortal." The conclusion, Socrates is a mortal, is entailed by the premises of the syllogism. So, too, the trajectories are entailed by integration of Newton's differential equations of motion and the initial and boundary conditions.

As discussed in Chapter 2, with Laplace and his demon who knew all the positions and momenta of all the particles in the universe, this became the birth of modern reductive materialism, and the view that the demon could calculate via Newton's laws the entire future and past of the universe. Laplace's dream that determinism necessarily implied predictability died with Poincaré's discovery of deterministic chaos in the case of three mutually gravitating objects, but determinism remained, prior to quantum mechanics.

The Laplacian view is central to us in that it expresses the "causal closure" of classical physics. Aristotle had said, "No first cause"; that is, no uncaused processes can occur. This holds in classical physics. Everything that happens has a cause, which became with Newton and Laplace, the causal *closure* of classical physics. Thus, ignoring for the moment outside influences, the behavior of the balls on the billiard table is entirely due to the forces between the balls and the initial and boundary conditions. There are no outside causes. If we allow gusts of wind to influence the balls, these classical causal inputs can be included in a fuller description of the billiard table and balls, but with the problem that we are now indefinite about the "boundaries" of the system. To handle this, physicists typically enclose a part of the universe as "the system," and deal with it, if classical physics, as causally closed.

The causal closure of classical physics leads directly to the "stalemate." Suppose the brain is a classical physical system of classical neurons, classical physics and chemistry. Then the current state of the brain is entirely sufficient, along with the classical physics dynamical laws of motion in differential equation form, to determine the next state of the brain and hence body. But then *there is nothing for conscious mind to do. Worse, there is way for the "mind" to "do it."*

Because of the causal closure of classical physics, its laws of motion have already included all the causes, and "mind" or "consciousness" has no causal role to play. Thus, in classical physics, there is nothing for mind to do and no way for mind to do it.

The idea that consciousness can play no causal role is deep in Western philosophy. Struggling with this and mind, Spinoza at the time of Descartes proposed a single substance with dual aspects, physical causal and mental, but no causal connection between them, rather a kind of God mediated harmony, as time flowed forward.

Leibnitz, in 1710, proposed his famous monadology, in which monads of three orders, entelechies, souls, and rational souls, existed, the latter two with consciousness. These mind-like monads do not interact causally at all, but exist with a pre-established harmony as time unfolds, and matter is somehow associated with the monads.

Spinoza and Leibnitz are, in part, efforts to deal with the causal closure of classical physics, and find mind, but based on the seventeenth century's idea of matter. Quantum mechanics takes us far beyond that seventeenth-century conception of matter and escapes the causal closure of classical physics! We can find an answer to Descartes' conundrum: There is nothing (causal) for mind to do and no way for mind to do it. We will find that a mind partially quantum and more can have *acausal* consequences for the meat of the "classical brain." Therefore, I will base my own approach to the subjective pole very much on quantum mechanics. In the ensuing chapters we will find an ontological basis for consciousness and free will and, beyond that ontological basis, a basis for our subjective world. Penrose (1989), discussing also his work with Hammerof (Penrose 1994), can reach similar conclusions on different grounds. I remark that much of the material in Part II appeared earlier in *Answering Descartes: Beyond Turing* (Kauffman 2011 and in press), and in "Beyond the Stalemate: Mind-Body, Quantum Mechanics, Free Will, Possible Panpsychism, Possible Solution to the Quantum Enigma" (Kauffman 2014), a very long title of which I am not particularly proud.

Responsible Free Will and Doing: *Initial Comments*

Our subjective pole includes our ordinary sense that we sometimes act with responsible free will. Free will must be central to our discussion.

It is recognized by all that the determinism of classical physics seems to preclude responsible free will. I pick up a frying pan and kill the old man, but I am not responsible, for my behavior was entirely determined.

However, I wish to raise a further profound issue about what we normally take responsible free will choices to include.

"I decide to drive my car to the store, but I could have decided to ride my bike to the store," is a typical case of our ordinary considerations of choice. A deep feature of this concept is that I could have done something different, driven my car, rode my bike. But this implies that the *present could have been different!* I *could* have arrived at the store in my car, or I could have arrived on my bike. But then at the moment I arrive at the store, in one case my car is parked at the store, in the other, my bike is parked at the store. Thus, at that instant, the present could have been different. To say that the present could have been different is to allow that *counterfactual statements about the present can sometimes be true. My car or my bike could have been parked at the store at a given moment.*

In a classical physics world, due to the causal closure of classical physics, the present, given the same initial and boundary conditions, *cannot have been different.* Counterfactual statements about the present can, therefore, never be true, given the same initial and boundary conditions.

But superb philosophers, wishing to hold to classical physics, hope to find a sense of "can" and "could" that will suffice for our sense of free will (Dennett, personal communication 2015). I give with pleasure and admiration the analysis of philosopher Daniel Dennett, given at the Santa Fe Institute and on line, in 2014. Dennett begins with an example by philosopher John Austin, who wishes to consider the claim: "I could have sunk the (golf) putt." This means, say Austin and Dennett, that under a variety of similar circumstances, "I" typically do sink the putt. Thus, I am *competent* to sink the putt under these circumstances, so "I could have sunk the putt" means I am competent to do so in a diversity of circumstances. Dennett stresses the variety of circumstances. This, hopes Dennett, is an analysis of the troubling word "*could*," which seems to allow a counterfactual claim. But does it? Yes and no. What are we to make of the statement, "Dennett could have *not* sunk the putt." Does this claim merely mean that Dennett failed to sink the putt, his competence failing him? What about Dennett's choosing instead to whack the ball into the clubhouse through the window, startling people at their tables as the ball lands in some salad? We mean, I claim, by choice, and "could" the latter, not the former mere failure of competence. In short, we mean that we could have chosen to do otherwise and the present could have been different. In private conversation, at the Santa Cafe, a lovely restaurant in Santa Fe, Daniel further defended his approach. "Stuart, you are using mere Actualism." "Good grief, what is Actualism?" I asked Dan. Roughly he said, Actualism is the view, given determinism, that only what Actually happened is what can or could happen. This, said Dan, is inadequate, for were the initial and boundary conditions different, something different could have happened.

Well? Recall (and Dennett may have played a role in these ideas; if so, thank you!) in Chapter 2 I pointed out that in classical physics, the boundary conditions of the billiard table *do* support an important kind of counter factual claim: The rolling balls do not hit the boundary of the walls in more than a few places in a short time, but we use the boundary conditions in classical physics to claim that were the initial conditions different, balls in different positions and with different momenta, then Newton's laws would still hold and the balls *would have* had different trajectories, which we can also calculate. So counterfactuals are allowed to be true, but not from the *same initial conditions and boundary conditions*. Indeed, since we cannot test all initial conditions, the set of possible initial conditions and given boundary conditions create the very phase space of all *possible* positions and momenta of the balls. The possible enters classical physics here and via the noncausal ergodic hypothesis, as I claimed in Chapter 2.

Then what about Dennett's "Actualism"? Both Newtonian classical physics and general relativity are all Actualism. The becoming of the universe in
general relativity is a causal structure of what actually arises in space-time
as a directed graph among events given the initial state of the classical universe, that is, the mass distributions and initial space-time curvatures. In
just what sense in classical physics "could the initial and boundary conditions have been different"? Consider the deist God of the eighteenth century
who created the universe, initial and boundary conditions, and let Newton's
laws take over. Well, God could have created different initial and boundary
conditions. But that is truly a deus ex machinus, outside of classical physics
for the simple reason that the statements of the initial conditions are outside of Newton's laws of motion. Or: In classical physics in continuous space
and time or space-time, we use real numbers and the real line, which include
infinitely long decimals such as the nondenumerable infinite number of irrational numbers. We cannot know the initial conditions to infinite precision,
so if the system is chaotic, we cannot predict what will happen. Stating the
initial conditions fully seems impossible. But that failure does not alter that
idea that whatever those conditions may be, the world can only be the history
of Actuals. The case beyond Actuals is not helped if we suppose the universe
is finite on the Planck space scale, hence not describable by the real line. That
makes Actualism more rigid, not less in classical physics.

I conclude that if we restrict ourselves to classical physics, we cannot get
beyond Actualism, and that Austin, me, or Dan being capable of sinking the
putt is not what we mean at all by "could." We mean, I conclude, that counterfactuals can sometimes be true and at a minimum the present moment
could have been different; my car or my bike could at this moment have been
parked at the Safeway.

In general relativity "the present moment" is located as a dimension in
space-time. Standard general relativity (GR) has no way to allow "the present to have been different," for GR is entirely deterministic and is the graph
of connected Actual causal events in space-time. Penrose has proposed his
"orchestrated objective reduction" (Orch OR; Penrose 1989, 1994), as I will
discuss later, which modifies general relativity so space-time can "split" into
two or more possibilities, or quantum superpositions, that later "collapse
into only one of these that becomes actual by gravitational self-interaction.
Note that Penrose has no deductive mechanism for which "choice" the universe makes. So on Penrose, the "present," that is, the very structure of space-
time, could have been different, for only one of the two possibilities becomes
"actual" upon collapse with a burst of consciousness and will.

I shall discuss in Part II, in addition to Penrose, the potential role of quantum mechanisms in many ways for helping get us to our subjective pole. A first and central point is that on most, but not all, interpretations of quantum mechanics, as noted in Chapter 2, quantum measurement is both *real* and *ontologically indeterminate*. The electron *could have been measured to be spin up or measured to be spin down at a given moment*. If these many, but not all, interpretations of quantum mechanics are correct, then counterfactual statements about the present can sometimes be true. At the present moment, the electron could have been measured to be, and so is, spin up, or could have been measured, and so is, spin down. Thus, and critically, quantum mechanics on many interpretations does allow the present to have been different, a necessary condition, I claim, for a responsible free will able to truly choose.

Also, again, on the multiple-worlds interpretation of quantum mechanics, the universe splits at each "measurement," where measurement is not real, but the knowledge or consciousness of the physicist and the world splits at each "measurement." But since measurement is not real, in no single one of the branching universes could the present have been different. In a sense, however, on multiple worlds, the present could have been different, but only in now different universes.

For completeness I note again that on Bohm's interpretation of quantum mechanics, the outcome of measurement is deterministic. Many think his interpretation, entirely nonlocal, is not acceptable. His quantum potential plays a noncausal role in "guiding" particles deterministically but nonlocally. To my knowledge, he offers no account of nonlocality, so deep and pervasive in his theory.

I have commented at some length about classical physics, determinism, and responsible free will, and I will turn to quantum mechanics as a major theme in the subsequent chapters as the only way I now see based on current science to break the causal closure of classical physics. Breaking that causal closure is necessary not only to hope for a responsible free will but also to avoid a purely witnessing, or "epiphenomenal" mind, which *is* possible in classical physics, my next issue.

Classical Physics Allows an Epiphenomenal Conscious Mind

The causal closure of classical physics, in which there is nothing for mind to do and no way for mind to do it, does not preclude a consciousness associated with a classical physics brain and body. However, that consciousness can only "witness" the world; it cannot alter or affect any changes in the classical physics world. Such a mind is called "epiphenomenal." In classical physics, we can have an epiphenomenal conscious mind. That mind cannot alter the becoming of the actual classical physics world, so of course it cannot have responsible free will by altering the world, and in the sense discussed earlier in which the present could have been different.

If mind is epiphenomenal, then why did our complex brain and conscious mind evolve? An epiphenomenal mind cannot help the tiger chase the gazelle. One answer could be that consciousness enchants us and so is of selective value even if epiphenomenal. Well, ok, but I hope we can do a lot better with real free will and a consciousness that can alter the becoming of the universe in part via that responsible free will and doings. This is just what we think of our own subjective pole. To reach this in Part II, I will suggest that the same is true of the universe, a panpsychism including electrons or protons exchanging photons, or perhaps fermions exchanging bosons. None of this is in standard quantum mechanics but hinted at by the quantum enigma (Rosenblum and Kuttner 2006), a place where, to the discomfiture of many physicists, physics seems to point beyond itself and toward the subjective pole.

Bertrand Russell, the Early and Late Wittgenstein, Then Artificial Intelligence

If classical physics allows only an epiphenomenal mind that can observe but not act, that has been fine with British empiricist philosophers from Hume onward. These philosophers have focused on how we can reliably *know* the world and have ignored entirely *acting* in the world. I take as my favorite example Bertrand Russell (1956), who in the early twentieth century wished to find the most reliable source of knowledge of the world. We could be wrong that there is a chair in the room, but we are far less likely to be wrong that we "seem to see a chair in the room." That is, our awareness of our "sense impressions" is less subject to error, despite hallucinations, than claims about the real world. Now look about you. You see an integrated visual world, the "unity of consciousness," which will concern us later. But Russell, whole cloth, shattered this unity of consciousness by proposing that we are aware of bit of sense data such as "red here" or "A flat now." Then he wished to ascribe "sense data statements," true or false: "Kauffman is now aware of A flat." Then the hope was to use the first-order predicate calculus, with quantifiers such as "All" and "At least one," and yes no values of sense data statements to derive the external world. The effort, starting with *Principia Mathematica*

(Whitehead and Russell 1925), which reached its climax in Wittgenstein's Ph.D. thesis, *Tractatus Logico-Philosophicus* (Wittgenstein 1921), failed. No set of sense date statements is logically equivalent to "There is a wing chair in the room." In philosophy, logical reduction or logical equivalence requires that a sentence in a higher level language can be replaced without loss of meaning by a prestated set of propositional statements linked by logical calculus, for example, in the lower level language.

Note that for Russell and the early Wittgenstein in *Tractatus*, the concern is with knowing the world, not acting in it. So a merely epiphenomenal mind is not of central concern and classical physics will do. In it we will be either zombies or have merely an epiphenomenal mind.

Russell's effort reached its culmination in Wittgenstein's *Tractatus*. Wittgenstein later abandoned the entire effort in his transformative "Philosophic Investigations" (Wittgenstein [1953] 2001). Consider what he called "language games." For example, consider "legal language games." For example, "Kauffman was found guilty of murder on August 7, 2014, in Superior Court, Sacramento." To understand this statement, said Wittgenstein, requires a co-defining *circle of concepts*, including, law, breaking the law, guilt, innocence, trial by jury, habeas corpus, legal capacity to stand trial, legal admissibility of evidence, and a wide range of other concepts. We *learn* these concepts. But they constitute an irreducible "language game," even a "way of life" for Wittgenstein in the *Investigations*. We cannot, he claims, reduce legal language game talk to talk about ordinary person actions, nor these to physics mass points, sound waves and light waves. In short, philosophers have given up on the idea of a basement language to which all we say can be reduced. Yet artificial intelligence, ignoring the Wittgenstein warning about nonreducibility, flourishes and is, among other things, to date, incapable of crossing among irreducible language games. Another point, if there is no basement language, including quantum mechanics and general relativity, reductive materialism will not get us to legal language games where I can, in fact, be found guilty and physically hung.

In 1943, Warren McCulloch and Walter Pitts published a seminal paper that founded modern artificial intelligence, "The logical calculus of ideas immanent in the mind" (115–133). These authors invented "formal neurons" or logic gates with excitatory and inhibitory inputs which would "fire," that is, turn from off (0) to on (1), if the "weights" of excitatory inputs minus the weights of inhibitory inputs exceeded a preset threshold. The authors invented an M by N matrix of rows of binary variables, light bulbs, N in each row, M rows deep. Formal neurons made connections, depicted by arrows, from themselves to formal neurons in the *next* of the M rows, leading from the input row, the first of the M rows, to the last of the M rows. Lacking feedback loops, this is a "feedforward" formal neural network. McCulloch and Pitts showed that such a network could compute any "logical" function, called Boolean functions, taking values of only 1 and 0, on the final row, given any list or vector of 1 and 0 values present at time 1 in the first row. In the first model, time comes in discrete moments at which all neurons update on their inputs simultaneously. There is a finite number, 2 raised to the 2 raised to the K, Boolean functions on a Boolean light bulb variable with K inputs, so the set of functions for this network is finite for any fixed network and logic.

The McCulloch and Pitts paper was seminal and has been generalized to networks with feedback loops and with all possible Boolean functions, not just excitatory/inhibitory, or "threshold" Boolean functions. I will discuss some of the aspects of these networks in later chapters for they bears on reliable behavior that, at least in the classical world, must be part of how systems can "know their worlds, evaluate the world as to whether the state of the world is 'good or bad' for the agent, and act reliably" on their worlds (Kauffman 1993, 1995). But to "act" with choices where the present can have been different, we will need to go beyond classical physics to get beyond epiphenomenal mind.

How did "mind" enter this network at all? Implicitly, McCulloch and Pitts imported the true-false values of Russell's sense data statements; "red here" is true for light bulb one, into their calculus. But, of course, this importation is not justified at all. All the formal neural network is, is a set of, say, electronic gates turning one another on and off in a subset of classical physics allowing only discrete (0,1) states, and discrete time T, T + 1. These logic gates are purely syntactic; they have no semantics. Were such gates to carry actual "sense data," that consciously available sense data would constitute semantic content.

Despite being merely syntactic, a vast amount of work has been done deriving from McCulloch and Pitts. One line of work is the Hopfield model, in which variables are again discrete, here −1 and +1, and connected to one another in specific ways. Each variable updates continuously or discretely in time to minimize an "energy" function, like a classical ball rolling down a hill, called a Hamiltonian. Put simply, this Hamiltonian is arranged to have a number of valleys with minimal energies at their bottoms. The state space of the Hopfield model is all the possible initial values of the −1 and +1 variables each at some point on the Hamiltonian landscape, and those initial states varying only in the value of a single variable (e.g. changing from −1 to +1 being near one another on the hilly landscape). From each initial state, the system follows a trajectory, the analogue of the billiard balls, flowing down to the bottom of the valley it starts in and coming to rest at a steady state. The minimum is called the "memory" of the initial state, and the set of states in the valley following trajectories to the same steady state is called the "basin of attraction of the minimum," which is also called "the attractor" because it "attracts" the trajectories that flow into it. The basin is considered the generalization of the memory, and it gives a "content addressable memory" for if the system is started anywhere in the same valley, it will flow to the same single steady-state attractor memory at the bottom of the valley. Hopfield nets are wonderful and by tuning the connections and energies between the variables, the landscape, its valleys and their minima, or attractor/memories can be sculpted in many ways. Much work has been done on the expected number of minima or memories at the bottoms of valleys, the old result being .14 n for n binary variables, hence .14 n memories. But note that the Hopfield model cannot solve the nonreducibility of language games issue raised by Wittgenstein. The Hopfield network is also merely syntactic, not semantic, and thus not even a candidate even to "play" different nonreducible semantic language games.

Another fine technique for tuning couplings between discrete or continuous variables uses "back-propagation," a way of changing the "weights" in the couplings, so the system approaches a mapping from a desired input state to a desired output state in feedforward networks. Back-propagation is a kind of generalization in a feedforward network of the first McCulloch and Pitts feedforward network, both yielding a mapping of input to output states. Much work has gone into "training" by supervised learning such networks to "classify" sets of inputs into "desired" output classes, one form of current machine learning.

All this is wonderful, amazing work, which is part of how your telephone helps find telephone numbers. In this huge area of computer science, words are represented as binary strings and word linkage maps are made and stored, and searched by universal computers, that is, our computers. Word maps are a huge field. But the "words" are merely syntactic, with no semantics whatsoever. Nevertheless, the hope is by syntactic maps between syntactic words, one can capture real language. So far, this has failed. A salient example of failure is: "Time flies like an arrow," a metaphor that requires semantics to make sense of it and is hence hard or impossible to capture in syntactic word webs. A second failure is the "screwdriver" argument, listing all uses of screwdrivers or finding a new use of a screwdriver is not achievable algorithmically. This is the famous "frame problem" of computer science.

But all the aforementioned artificial intelligence ignores entirely Wittgenstein and others point that we cannot reduce a higher level, for example legal language (game) to lower levels such physical descriptions. If we cannot so reduce legal talk to physical talk, we should be deeply worried about reductive materialism, that some law down there entails all that arises in the universe. If we cannot reduce legal language, such as "Kauffman committed murder," whose understanding requires a web of other legal concepts such as "guilty, able to stand trial, admissible evidence, law itself, breaking the law," and so forth, to ordinary human actions, and then to physics, then no physics will allow the inverse, to infer the meaning of "Kauffman was found guilty of murder" or the consequent fact that I am physically sent to jail for forty years without parole possibilities for the first thirty years. This is a physical change in the universe, not just words. Yet we all understand these legal statements perfectly well and my being sent to jail. This language game issue seems unapproachable by propositional-based artificial intelligence. And, of course, our legal talk assumes a sometimes more than epiphenomenal conscious mind sometimes capable of responsible free will. Both are precluded on classical physics.

A new area of artificial intelligence is emerging called "embodied mind." Here is an example: A colleague has a robot with a cousin of a formal neural network brain, and a body with sensors that acts in the world. By mutating the connections and logic of the neural network brain, the robot does in fact evolve to perform new "functions." Nothing in these robots is "propositional" like linked if-then statements, followed by "*do if.*" *Absent* total reliance on propositions, the frame problem, classical systems seem able to evolve new "functions" in a sense worth exploring, although the present could not have been different. There seems to be an important analogy between such "embodied classical physics mind" and its capacity to evolve and Darwinian evolution and Darwinian preadaptations, which may not be prestatable, but the latter is beyond current work. If we are willing to use the word "function" for a robot that "solves" a task by selection, prestated or later not even prestated perhaps, then perhaps we have a rudiment of semantics in the notion of "function" itself, with no need to evoke consciousness yet. Note again that such robots are classical physics and cannot "choose" in the sense that the present could have been different. Nor is it clear that they can know semantically nor solve the language game issues. Nor does it seem that such systems, were they conscious, could be other than epiphenomenal mind, unable to alter the world.

The Turing Machine and Algorithms

Almost all of you are familiar with Turing, and algorithms, famously done on the Turing machine (Turing 1968), which became the basis of the universal computer, including the one I am typing on.

Turing and many have hoped that the Turing machine was an adequate model of mind, for example as tested by the Turing test (Turing 1968). I think it fails for several reasons: (1) There is no coherent argument to associate consciousness with a Turing machine, any more than Russell's sense data can be imported into the McCulloch Pitts formal neural net. (2) Were Turing machines conscious, that consciousness would be epiphenomenal at best. (3) As a subset of classical physics, the Turing machine cannot support counterfactual claims that the present could have been different, so it cannot support a responsible free will, as discussed. (4) Turing machines via good old-fashioned artificial intelligence based on true and false propositions cannot solve the famous "frame problem," which is identical with the "uses of the screwdriver" argument of Chapter 4. The number of uses of a screwdriver is both indefinite and unorderable, so no algorithm can list them all or find new uses, which uses are merely a list or nominal scale. This argument says for evolution and the human mind (recall the Tokyo man's business venture from Chapter 5), neither evolution nor the human mind can be merely algorithmic. If so, we are not propositional Turing machines. Embodied mind robots on a computer-Turing machine may be able to evolve new "functions" if we are willing to use the word "function," but they cannot have free will in the sense of counterfactual statements about the present. (5) We should be deeply suspicious. In Chapter 1, I pointed out that we use metaphors all the time to guide us and that language may have started metaphorical (or even gestural) and then become propositional. But metaphors are neither true nor false! Yet we are evoked by them and use them all the time. Why are we limiting ourselves to minds using only true-false propositions as how we know and are in the world? Indeed, a metamathematical issue: I think we would agree that no prestated set of propositions can exhaust the meanings of a metaphor. If not, metaphors are richer than prestatable in propositions, and we use them. Furthermore, as I noted, if by a mathematical proof, we mean an entailing set of true-false propositions, then no set of propositions can "prove" that no prestated set of propositions can exhaust the meanings of a metaphor. We understand this "proof," but it is not mathematizable. Penrose, in *The Emperor's New Mind* (1989), made similar arguments decades ago about understanding Godel's incompleteness theorem. The human mind can understand what cannot be done algorithmically. The mind seems to be richer than algorithms and mathematics itself. I take this argument seriously. I here claim that no mathematics can prove that "no prestated set of propositions can exhaust the meanings of a metaphor." This is, of course, another negative result, like the "uses of the screwdriver" argument. Neither can be "mathematized" in the sense of listing all the meanings of a metaphor with any prestated list of propositions, or listing all the uses, or next use, of a screwdriver. In a trivial sense, of course, merely saying that this cannot be stated is a kind of "metamathematical" proof, but that point does not deny the seeming validity of my arguments. If the claim about metaphors is true, like Wittgenstein's language games (1953), mind is not algorithmic. Embodied mind robots may, however, turn out to be able to evolve new "functions," perhaps even that cannot be prestated if the new functions can be "used" without prespecification of the "function." This does not seem impossible, and is a new form of artificial intelligence very much worth pursuing, but there is no reason to think the robot conscious and were it conscious, that consciousness must remain epiphenomenal. And such embodied mind robots cannot choose in the sense in which this requires that the present could have been different, given the same initial and boundary conditions, here including the same program and random seeds in the program.

Conclusions

I have tried to sketch briefly some of the attempts using classical physics and subsets of it in Turing machine algorithms, neural network models, continuous and discrete in time and state, with fine models of memory and generalization, to show that classical physics cannot yield a nonepiphenomenal consciousness, nor can it yield a counterfactually different present; hence, it has no hope to allow for a responsible free will that can have chosen differently. Spinoza, Leibnitz, and others using the seventeenth-century concept of matter attempted kinds of panpsychism, but such efforts seem to fail. In the rest of Part II, I build toward a potential set of answers, most testable, most reasonable. Quantum mechanics allows us to break the causal closure of classical physics.

7

A New Proposed Dualism

RES POTENTIA AND RES EXTENSA, LINKED BY QUANTUM MEASUREMENT

I BEGIN MY multichaptered discussion of quantum mechanics and its potential role, including the quantum enigma, if it is real, in offering a path not only beyond the causal closure of classical physics that demands at best an epiphenomenal mind, but toward a more than epiphenomenal mind and the possibility of real responsible free will in which the present could have been different. My discussion takes many steps, and I hope for patience from the reader. The present chapter proposes a new dualism, res potentia and res extensa, ontologically real Possibles and real Actuals, linked by measurement. This proposed dualism is part of a new, hopefully consistent, interpretation of quantum mechanics and also includes *Mind* in the triad, Actuals, Possibles, and Mind, where Mind observing and choosing, acausally converts Possibles to new Actuals, which acausally create new Possibles for Mind to observe to create yet new Actuals in a continuous becoming, among quantum variables, upward to us. The dualism, not yet including "mind," has roles in the larger context of Part II, which are as follows: First, the new interpretation is one of two lines of argument that probably, but not surely, rule out a "deductive mechanism" either for quantum measurement or for the "choice" made at measurement. Furthermore, the absence of such a mechanism is testable by finding a mechanism for measurement. No such mechanism has been found since the Schrodinger equation and Born interpretation in 1927 and 1928, but Penrose's orchestrated objective reduction (1989, 1994) is a proposed mechanism and testable, although as noted it provides no deductive mechanism for the "choice" between multiple superpositions of space-time that he proposes.

A second larger potential role for the claims of the present chapter concerns Einstein, Podolsky, and Rosen (1935) and verified nonlocality for quantum entangled particles. We have no way that I know of to think about this nonlocality that is at all acceptable. Res potentia, ontologically real Possibles, appears to offer such a pathway. Res potentia may also afford an answer to the puzzle that in quantum mechanics, if measurement is real, measurement instantaneously changes the wave function. If instantaneous, it cannot be causal, for casual influences cannot travel faster than the speed of light. In addition, ontologically real Possibles afford an answer to two other deep puzzles about quantum mechanics discussed later.

The third role is linked to the quantum enigma (Rosenblum and Kuttner 2006), discussed in Chapter 12, where physics seems to point beyond physics. The quantum enigma itself, if true, also points to at least our free will in which the present could have been different, and our consciousness can testably be shown to be sometimes sufficient for quantum measurement. If shown true, consciousness is real. Then this aspect of our subjective pole is real. A proposed solution to the enigma is my proposed Triad: Actuals, Possibles, and Mind. In this chapter I discuss only Actual and Possibles, and in Chapter 12 I discuss Mind.

I begin with straightforward quantum superpositions. The Schrodinger equation is a linear wave equation that propagates waves in time and space. Critically, there is *no energy* term in this equation, so whatever is "waving" is not matter or energy. No one knows what is waving. I will propose, perhaps with Heisenberg well before me (1958), and Zeh (2007), that "possibilities that are ontologically real are what are "waving." A quantum system can be prepared in a superposition of states, say the electron has a wave amplitude to be measured to be spin up, which amplitude when squared, according to the Born rule, yields a probability of 50 percent up, and a superposition probability of being measured to be spin down of 50 percent.

It is essential that this superposition is *simultaneous*. The famous example is the Schrodinger cat paradox in which a cat is prepared in a superposition of dead and alive, and will die if a quantum random radioactive decay occurs, but remains in this simultaneous state of *both* dead and alive until the box with the cat in it is opened and by being measured the cat is either dead or alive, not both.

The central point here is that the superposition state entirely violates Aristotle's "law of the excluded middle." Here is an example: "The cat is on the mat" or "The cat is not on the mat." There is nothing in the "middle," so "The cat both is and is not simultaneously on the mat" is a contradiction.

Quantum superpositions, in general, and in particular for entangled particles, violate Aristotle's law of the excluded middle.

A truly astonishing feature of measurement, that is, here looking in the box and seeing if the cat is dead or alive, *if* measurement is real, is that after measurement, the cat *is* either dead or *is* alive, not both. So *after* measurement, Aristotle's law of the excluded middle is fully obeyed. Measurement, if real, takes a system that does not obey the law of the excluded middle to a state that does obey the law of the excluded middle. Indeed, the failure of the superposition "state" to obey the law of the excluded middle is exactly why physicists say we cannot assign a "real state" to the superposition, until we get the result of measurement, when a state, true or false, *can* be assigned to the quantum system.

Feynman's Formulation of Quantum Mechanics as a Sum over All Possible Histories

Nobel Laureate Richard Feynman formulated a sum over all possible histories (Feynman, Leighton, and Sands [1963] 2006, 2010), a consistent version of quantum mechanics in which, say in the famous two-slit experiment, *each* electron *simultaneously takes all possible paths through space and the two slits to the recording film screen beyond it*, which later reveals the interference pattern of light and dark bands on the developed film as stable spots, one per measurement. But this formulation must say that a single electron simultaneously does and does *not* go through the left slit. But that statement, like the former superposition example, does *not* obey Aristotle's law of the excluded middle. After measurement, in which a spot appears at a specific point on the film, the spot is or is not at that point, so again now it does obey the law of the excluded middle. Once again, before measurement the coherent quantum system does not obey the law of the excluded middle, but after measurement, the outcome does obey that law.

Another outcome of quantum measurement of a superposition can be a single specific wave function from among the superpositions. That single wave function, still quantum, at the moment of measurement also obeys Aristotle's law of the excluded middle. Later in time that single wave function, quadratically in time, flowers further wave functions. Outcomes of measurement do obey the law of the excluded middle, whether spots on a screen or single wave functions. Quantum superpositions do not obey the law of the excluded middle.

The next point was made by philosopher C. S. Peirce (Kauffman 2012) in the early twentieth century. Actuals and Probables, he said, do obey the law of the excluded middle. For example, the cat is or is not on the mat. Or the probability of 5,460 heads out of 10,000 flips of a fair coin is .039, on the frequency interpretation of probability, can be tested by flipping a coin 10,000 times, counting the number of heads, and repeating this experiment a million times to get a distribution of the number of heads. If the probability statement is *true*, the mean will hover near .039. So the statement "The probability of 5,460 heads out of 10,000 flips is or is not .039 with a tiny standard deviation" is true or false and fits the law of the excluded middle. Probability statements obey the law of the excluded middle. On a Bayesian interpretation of probability, the same remains true, given a *prestated* set of outcomes, and the priors, we can calculate the posterior and use that to update our priors. But we knew all the possible outcomes ahead of time and could count them; hence, the probability statement is true or not, and if not, we update our priors by Bayes rule to converge on the true probability.

Probabilities, as Peirce said, obey the law of the excluded middle (Kauffman 2012).

But, said Peirce, *possibles* do *not* obey the law of the excluded middle. Thus, "The photon *possibly did and simultaneously possibly did not go through the left slit*" is *not* a contradiction.

I now want to take a simple but very big step and propose a new dualism in the universe, "res potentia, ontologically *real* possibilities, and res extensa, ontologically real Actuals, linked by measurement. Unlike res cogitans and res extensa of Descartes, which is a *substance dualism*, this proposed dualism is *not* a substance dualism as possibilities are not substances precisely because they do not obey the law of the excluded middle. By definition, "Actuals" obey the law of the exclude middle, whether a single wave function as a result of a nondestructive measurement or a measurement yielding a stable spot on a silver halide screen, somehow "classical," where accounts of "classicality" differ. I will define Actuals as located in space and time up to special relativity. Thus, I propose that the spot on the silver halide screen *is* in a "small region" in space and time up to special relativity. The single wave function just after measurement is also localized in space and time up to special relativity at the moment of measurement, but then it spreads out and also flowers new amplitudes quadratically in time. Both the spot and the specific single wave function obey Aristotle's law of the excluded middle, and so they are Actuals by my definition.

The hypothesis that *possibilities* are ontologically real is a very big step for us. We have thought for millennia that only Actuals are ontologically real. Indeed, general relativity is precisely a theory in which the entire world is a four-dimensional space-time manifold of actual events which stand in causal relation to one another. The structure of this web of causality *is* reality. There are *no* Possibles in Newton or special or general relativity; nor are there in classical physics in general.

With this in mind, is it a crazy hypothesis that Possibilities are ontologically real? I think it is not.

First, I want to note again that one of the founders of quantum mechanics, Werner Heisenberg (1958), had much the same idea many years ago. He thought of quantum coherent behavior as "potentia." Here is his statement: Heisenberg as quoted in *The Quantum Enigma* (Kuttner 2006, p. 104): "In the experiments about atomic events we have to do with things and facts, the phenomena are just as real as any phenomena in daily life. But the atoms or elementary particles themselves are not real; they form a world of potentialities or possibilities rather than one of the things of facts."

Next I quote physicist Dieter Zeh (2007) at length:

> in classical physics you can and do assume that only one of the possibilities is real (that is why you call them possibilities). It is your knowledge that was incomplete before the observation. Mere possibilities cannot interfere with one another to give effects in reality. In particular, if you would use the dynamical laws to trace back in time the improved information about the real state, you would also get improved knowledge about the past. This is different in quantum theory (for pure states): In order to obtain the correct state in the past (that may have been recorded in a previous measurement), you need all apparent "possibilities" (all components of the wave function—including the non-observed ones). So they must have equally been real.

Heisenberg, Zeh, and I are pointing to what I am calling ontologically real possibilities, res potentia. Yet as noted, potentia do not obey Aristotle's law of the excluded middle. But then after measurement, we have Actuals that do obey the law of the excluded middle. I claim that this is consistent with res potentia and res extensa linked by measurement, which converts Possibles to Actuals.

No less a thinker than Aristotle toyed with the reality of potentia. Alfred North Whitehead in his *Process and Reality* (1927–1928), constructed an ontology not of things and their relations, but of events that became "actual occasions" with a protoconscious "prehension." Possibles become Actuals become Possibles here. My own interpretation of quantum mechanics, here and later, may be similar to Whitehead's thoughts. More recently, Abner Shimony (Penrose, Shimony, Cartwright, and Hawking 1997, pp. 150–151) discusses quantum mechanics and the actualization of potentialities in measurement in a modernized Whitheadean approach. Epperson (2004) has also explored possibles in the relation between quantum mechanics and Whitehead. On this proposed new dualism, res potentia and res extensa linked by measurement, quantum mechanics is about this dualism, in which the world consists of both possibles and actuals, not just actuals, linked by measurement. This is a radical view, and stands in contrast to the epistemic view in which quantum mechanics is not about "reality," but what we can know of reality, the "epistemic" view held by Bohr in the Copenhagen interpretation of quantum mechanics.

New Actuals Can Acausally and Instantaneously Enable New Possibles

Central to the view I will develop is that new Actuals that obey the law of the excluded middle can acausally and instantaneously enable new Possibles. Then if measurement is real, its outcomes, which are new Actuals, can engender new quantum Possibles. For this, I claim, we need not invoke the "epistemic" view of quantum mechanics.

The evidence so far that I bring for res potentia derives from the failure of quantum coherent behavior to obey Aristotle's law of the excluded middle coupled with Peirce. But in Chapter 4 I discussed almost what I discuss here. Consider again the unprestatable emergence of Darwinian preadaptations such as the swim bladder, derived (think paleontologists) from the lungs of lung fish. Once the swim bladder exists as an Actual in the evolving biosphere, I suggested that that Actual is an "enabling constraint" that does not cause but enables a new Adjacent Possible empty niche, for a worm or bacterial species, but not a giraffe, might or might not evolve to live in the swim bladder. I think we accept this as true. I raise two issues, mentioned earlier: (1) Once the swim bladder exists in the evolving biosphere, is it true or not that the swim bladder now creates and enables a new Adjacent Possible empty niche? Yes. (2) Does the swim bladder *cause* the worm or bacterial species to evolve to live in the swim bladder? No! The relationship between the now actual swim bladder *as* the new niche is one of enablement. What then happens is, sometimes, quantum events that are sometimes quantum random mutations altering the DNA or RNA of the species. If quantum outcomes are acausal, and no mechanism for the choice of measurement outcomes is known, and if measurement is real, then the mutation arises acausally. The mutated DNA may yield new modified screwdrivers in bacteria or worms, that may be of selective value at the level of the entire worm or bacterial organism(s) living in the world of the swim bladder, so be selected at the level of the Kantian whole organism. Then the swim bladder does not cause but *enables—makes possible—the evolution of the worm or bacterial species, but not the giraffe, to evolve and adapt to live in the swim bladder*. But enablement is not a "causal" category; it is a "making possible." Similarly, the invention of the iPad and its arrival in Tokyo, given the prior existence of many crowded apartments in crowded Tokyo, a new *actual* that *acausally enabled, or made possible*, the new business of the Japanese man who realized he could scan the books of others living in crowded apartments into *their* iPads, sell *their* books, and take 5 percent of the sales price as his profit. The Actual existence of crowded apartments, books, and the arrival of iPads did not cause but enabled this new business opportunity to become *possible*, and the Japanese man seized the opportunity. The conditions of the iPad, the crowded apartments, and the books in them did not *cause* the new business; they afforded or *enabled or made possible the opportunity for the new business*.

What shall we say of the swim bladder as a new Actual *enabling* a new Adjacent Possible niche *opportunity or possibility* for the worm or bacterium but not the giraffe? Is the opportunity, or possibility, *real*? One is driven to say *yes*. But in the world of classical Newtonian physics, in which only the history of Actuals was what came about, the notion I wish for Possibles would make no sense. For Possibles in res potentia to make sense, at least a necessary condition again is that the present could, counterfactually, have been different. But if quantum measurement is *real* and ontologically indeterminate, as it is on most interpretations of quantum mechanics, then the present could have been different. The electron could have been measured to be spin up or spin down. (For the mutations in biological evolution of Chapter 4, I wish to say again that the mutation, hence that present moment when the mutation occurred, could have been different if the mutation is a quantum event. But the present at that moment could have been different: The mutation might not have occurred. This, I claim, is consistent with res potentia, ontologically real possibilities, which in the evolutionary case become opportunities for the worm to evolve to live in the swim bladder. New Actuals enable new Possibles.

Penrose, in *The Large, the Small and the Human Mind* (Penrose et al. 1997), discusses counterfactual "null" measurements with actual consequences in the world, experimentally demonstrated. So quantum mechanics does allow, as verified, counterfactual results of "null" measurements. I explore null measurements later in this chapter.

How "Fast" Do Possibilities Change?

Once the swim bladder is a *new* Actual in the biosphere, how long does it take for it to become *possible* for a worm or bacterial species, but not a giraffe, to evolve to live in the swim bladder? The answer seems to include "instantaneously," if worms or bacteria or other species that may seize the new niche opportunity already exist. Here we need not consider consciousness. But what about the man in Tokyo? Give the actuals of crowded apartments with books in them and the invention and sale of the iPad in Japan, how long did his new business opportunity, or possibility, take to become, well, possible? The answer is, the new possibility can have come to exist instantaneously or nearly so, given the new Actual presence of a single or more new Actual iPads in Tokyo and Actual presence of crowded apartments filled with books, two Actuals, one new. Thus, it seems possibilities can change rapidly or even instantaneously, when new Actual arise. How can something become *possible* instantaneously? The answer is that becoming Possible, enablement, is not a causal process, so possibilities can change instantaneously and acausally.

I give a further example: You and I plan to meet at the store on J Street tomorrow to buy orange juice and have coffee. Today, as I drive by the store, a notice is just going up: "Store closed immediately!" What just happened to the possibility that you and I will meet tomorrow at the store on J Street to buy orange juice and have coffee? The possibility *vanished*, and did so *both instantaneously and also acausally* the moment the store closed! Furthermore, all sorts of possibilities vanished acausally, for example, that Bill could drive there in a car the day after tomorrow and park for an hour and eat an ice cream cone. So if we accept that possibilities are *real*, then changing an *actual* can instantaneously and acausally change an indefinite set of possibilities. A new Actual acausally and instantaneously changes what is possible, eliminating some possibilities, making others now possible. For example, a new store might open at some moment on K Street, and now we can meet there sometime thereafter. A new Actual acausally and instantaneously creates new possibilities.

Furthermore, I claim, what is now really possible does *not* depend upon our knowledge, the "epistemic view of quantum mechanics" supported by Bohr. If the store on J Street closed at a specific moment T, today, then whether you or I know it, our possibility of meeting at the store on J Street thereafter vanished at time T. If we try to meet at the store after time T, we will "learn" (come to know) that we were wrong, even if, not knowing, we thought it was still "possible" to meet at the store after it had, in fact, closed. This happens in real life all the time. Of course, this claim depends upon something like a classical world, the store and our cars, and our free-willed choices of actions among possibilities in that somehow classical world, all discussed later in this chapter.

In quantum mechanics, if measurement is real and indeterminate, measurement can yield a single wave function, which is now a new Actual obeying the law of the excluded middle. But that new Actual then flowers quadratically in time new quantum amplitudes, so the new Actual single wave function does give rise to new quantum possibilities. Or measurement can yield a stable spot on the silver halide screen. That spot is a new "classical" actual due to measurement, and it does create new possibilities, for light bouncing off the spot behaves differently than were the spot not there. I discuss this in detail in Chapter 13.

Are Possibilities Located in Space or in Time?

Where in space is the possibility that you will fall on your head in the next moments? I will consider two ideas: (1) Possibilities are not located in space but in time up to special relativity. An interesting feature of this postulate is that ontologically real possibilities could have predated the Big Bang and lead to new ideas about the origin of the universe emerging from "the Possible." (2) Possibilities are located in both all of space and time up to special relativity. The view that possibles exist throughout all of space and time up to special relativity is supported by the standard view of normalizable probability distributions. Penrose (2004) makes this point: normalized probability distributions hold throughout all of space. Penrose states that only normalizable probability distributions can be physically real. But, I add to Penrose, nothing can be "probable," as in a probability distribution, if it is not first "possible." Thus physically real normalizable probability distributions throughout all of space must, I claim, be underlain by real possibles throughout all of space. I note that possibilities do seem to be located in time. First, in quantum mechanics, when a measurement occurs, wave functions change instantaneously, hence change in time, a mystery I return to in a moment. Second, a possibility coming into existence acausally depends upon the coming into existence of one or more enabling new Actuals. In a moment I will use this idea to explain why wave functions change instantaneously, hence in time, upon measurements that produces new Actuals.

But first I note that the postulate of the new dualism, res potentia and res extensa, linked by measurement, in which the world is both ontologically real possibles and ontologically real actuals, does not accord with "realism," where "realism" is concerned only the "actuals" of res extensa. Recent theoretical and experimental work (Gröblacher et al. 2007) casts strong doubt on "realism," claiming that realism must be abandoned, along with locality.

The postulate of ontologically *real* Possibilities affords a new and perhaps the best explanation for four mysteries in quantum mechanics: (1) nonlocality; (2) the instantaneous change in wave functions upon measurement; (3) "which-way" information; and (4) null "measurement" occurring where no direct measurement occurred. All interpretations of quantum mechanics seem to face these issues, even Bohm.

Nonlocality

The EPR paradox in 1935 has been described. Einstein, Podolsky, and Rosen realized that quantum mechanics implied an instantaneous nonlocal correlation for space-like separated but entangled particles upon measurement. This has now been confirmed repeatedly. The distance record is now 190 kilometers or so. The universe is nonlocal in the precise sense that *no causal influence* can travel faster than the speed of light. Hence, instantaneous nonlocal correlations cannot arise causally. As a specific case, if two entangled electrons can be measured to be spin up or spin down, by symmetry if one is measured to be spin up, the other can only be measured to be spin down. If, in fact, the first is measured "here" to be spin up, the second is instantaneously bound to be measured to be spin down.

Many workers struggling with nonlocality seek some form of "influence" that spreads faster than the speed of light (Stapp 2011). But it is also known that signals and information cannot be sent via nonlocality, so the influence cannot be so used. The nature of such "influences" is entirely unclear.

But if res potentia is ontologically real, possibilities throughout all of space can change instantaneously when Actuals change, and if measurement of one entangled particle yields a new Actual, that can instantaneously and acausally change the Possibles of the quantum coherent world that does not obey the law of the excluded middle, throughout all of space or perhaps even outside of space. Consider first the idea that possibilities are outside of space but inside of time up to special relativity. Altered Actuals acausally can instantaneously alter res potentia, Heisenberg's potentia, outside of at least space, and noncausally alter the results of measurement of the second entangled particle! In the specific case of two entangled electrons, there are, by symmetry, only two possible outcomes. Either the first is measured "up" or measured "down." Once it is measured, say "up," that is a new Actual and instantaneously and acausally, the *possibility* that the second will be measured "up" *vanishes*! The only possibility left is that the second will be measured "down." We have a framework to think about nonlocality via instantaneous *acausal* changes in res potentia, quantum coherent systems that fail to obey the law of the excluded middle and, by Peirce, can only be Possibles. (If Possibles are outside of space but inside of time, res potentia can have existed before the Big Bang, opening new ways to think about the origin of the universe, a topic I will briefly discuss in Chapter 14 with respect to Hawking's proposal of a boundryless universe (Hawking 2003), and Penrose's discussion of these ideas in his *Road to Reality* (Penrose 2004).

The common proposal in standard quantum mechanics is that physically real, normalizable wave functions are in all of space and in time up to special relativity. Then, as noted, on res potentia, if quantum behaviors are Possibles, those Possibles are not outside of space, but are inside of space, throughout all of space and in time up to special relativity. Then again, we explain nonlocality in the same way we do if Possibles are *not* in space but in time up to special relativity. Measurement of one entangled particle creates a new Actual that acausally and instantaneously alters what is possible. Here the possibility that the second electron is spin up has vanished, instantaneously and acausally, by measurement of the first electron as spin up; hence, it becomes "spin down." The alteration in possibles is acausal and instantaneous, so we can hope to explain nonlocality. We need not seek other faster than light "influences" that "condition probabilities." Altering possibilities suffices to alter, via the Born rule, probabilities.

Instantaneous Changes in Wave Functions Upon Measurement

As noted, for single independent particles in a superposition, and n entangled particles, measurement instantaneously changes the wave function. For a single variable in a superposition state, nondestructive measurement can yield the result that the variable is now in a single wave function, so fitting the law of the excluded middle and is now an Actual. Instantaneously, measurement has created an actual, a single wave function, but this new Actual, the single wave function, quadratically in time flowers new wave functions. Again, a new Actual enables new possibles instantaneously, even if the flowering of amplitudes is quadratic in time. (If the system is rapidly remeasured, it tends to be trapped in the initial wave function, the quantum Zeno effect I use later to obtain a "classical-enough world.") If n particles are entangled and one is measured, instantaneously the entangled wave function of the remaining $n - 1$ changes! If quantum coherent behaviors are possibles, and throughout all of space, a changing actual, the measurement of the first, can instantaneously and acausally change the possibilities, that is, the remaining wave function for the $n - 1$ entangled variables. An instantaneous change in wave functions cannot be causal. If possibilities are ontologically real, and quantum coherent behaviors that do not obey the law of the excluded middle are one or more possibilities, the latter in superpositions, are Possibles, we have a res potentia candidate explanation for the instantaneous change in wave functions upon measurement. To my knowledge, this mystery of quantum mechanics is otherwise unexplained, but in the mathematics of quantum mechanics, even in Bohm and Hiley (1993).

Clearly the ideas for nonlocality and the instantaneous change in wave functions are essentially the same.

Which-Way Information and Possibilities

In the famous two-slit experiment, if a measuring device is set up behind, say, the right slit to measure if the photon or electron passed, or did *not* pass through that slit, the interference pattern disappears. This is firmly established experimentally. But measurement of "which slit, or which way" the photon went, the right slit or *not* the right slit, is a *new actual* that can acausally and instantaneously change what is now possible. In particular, in this case, if the photon went through the right slit, that eliminates the possibility that it went through the one remaining, left, slit. That "left slit" possibility vanishes instantaneously and acausally when the new Actual, "Yes Right Slit," arose by "which-way" measurement. Conversely if the electron did *not* go through the right slit, that also is a new Actual that eliminates the possibility that the photon or electron went through the right slit, so that possibility vanishes instantaneously and acausally, and the only possibility remaining is that it went through the left slit, a null measurement. Again, a new Actual, Yes or No for the Right slit, instantaneously and acausally alters what is now possible. Some possibilities vanish in these cases upon measurement. With that vanishing, interference of possibilities vanishes as well. Which-way information eliminates interference.

In Chapter 12, I will discuss the quantum enigma. Part of it is that if an electron is prepared in a superposition simultaneously in box 1 and in box 2, and if we "look" in box 1 and "measure" that the electron is *not* in box 1, it is then measured to be in box 2, even though no one "looked" in box 2, again a null measurement. This is similar to which-way information. If the electron is observed to *not* be in box 1, that is a new Actual that eliminates the possibility that it is in box 1. This new Actual instantaneously and acausally leaves only the possibility that the electron is now in box 2, so the electron is now measured to *be* in box 2, and becomes a new Actual, although never "observed." This, again, is an example of a *null* measurement. Penrose, in Penrose et al. (1997), shows that null measurements can have actual consequences, counterfactually! Note that on these accounts of the four mysteries, we need *not* invoke the epistemic view of quantum mechanics, if possibles are ontologically real and measurement yields new actuals that then acausally and instantaneously enable new possibles.

All interpretations of quantum mechanics, including Bohm's vastly nonlocal theory with his nonlocal quantum potential, and multiple worlds, must explain nonlocality, the instantaneous change in wave functions, which-way information destroying interference patterns, and the Enigma's box 1 and box 2 issues. Ontologically real Possibles seem to offer a new explanation.

No Deductive Mechanism for Measurement

If res potentia, ontologically real Possibles, and res extensa, ontologically real Actuals, the former not obeying the law of the excluded middle for quantum superpositions, the latter obeying that law after measurement, is right and the two are linked by quantum measurement, there can be no deductive mechanism for measurement. The logic is very simple: "The X is *possible*" of res potentia does not logically *entail* the "X is *actual*" of res extensia. That is, if superpositions are two or more Possibilities, no deductive mechanism can "choose" among them in an entailed way to yield one Actual, that is, measurement and the "choice" made at measurement, if measurement is real. Thus, there can be no deductive mechanism for real measurement. This is not philosophy, but a testable scientific hypothesis, for it can be easily disproven were a deductive mechanism for measurement and the choice made ever found. None has been since quantum mechanics was invented in 1927.

In Chapter 12, I will discuss the quantum enigma (Rosenblum and Kuttner 2006) and propose that conscious and free-willed observation by humans and by quantum variables is necessary and sufficient for measurement. I will discuss new, very tentative evidence due to D. Radin (Radin et al. 2012, 2013) that human conscious attention, but not inattention or robots, can affect, at a distance, the measurement outcomes in the two-slit experiment. If confirmed—a big if—then if human consciousness can "mediate" measurement at a distance, quantum aspects of the human mind and quantum nonlocality for distant effects on measurement are the obvious hypotheses. Radin's experiments, even if confirmed, would not yet demonstrate nonlocality for the time interval between human attention a thousand miles away and alterations of measurement seem too long to demand nonlocality. (Radin [personal communication 2015] points out that if the participant were on Mars and the two slit device on Earth, positive results could establish non-locality, a doable experiment.) If we confirm that human conscious attention can alter measurement, we will have to consider the hypothesis that conscious observation even by quantum variables at the base of "classical systems," whatever "classical" may be, can mediate measurement, not our normal sense of measurement, and do so acausally, therefore, without a deductive mechanism for measurement. This will yield res potentia and res extensa linked by "mind," measurement in what I shall call The Triad: Actuals, Possibles, Mind, where Mind measures Possibles, creating acausally new Actuals that acausally create new Possibles for mind again to measure, creating new Actuals, then new Possibles for mind to again measure, in a persistent becoming, a status nascendi, somewhat similar to Whitehead in *Process and Reality* (1978).

Conclusions

I have presented a preliminary sketch of a new dualism, res potentia, ontologically real possibilities not obeying the law of the excluded middle, and res extensa, ontologically real actuals that obey the law of the excluded middle, linked, hence united, by measurement. This is not a substance dualism like res cogitans, thinking stuff, and res extensa, mechanical stuff, which seems to ask of mind to act *causally* on body, and then runs into the causal closure of classical physics, our stalemate since Newton. The Schrodinger equation has no term for energy, so whatever is "waving" is not matter or energy. And coherent behavior does not obey the law of the excluded middle. It is largely for this reason that we do not assign a "state" to a quantum coherent system. Not far from Heisenberg (1958) long before me, and Zeh (2007), I propose that quantum coherent behavior is propagating ontological real *possibilities* that do not obey the law of the excluded middle, res potentia. Measurement then converts Possibles to Actuals that do obey the law of the excluded middle. New Actuals acausally and instantaneously change what is now possible, so if Possibles are not located in space but perhaps in time up to special relativity, we have a proposal to account for nonlocality without seeking "influences" that somehow travel faster than the speed of light (and perhaps a new way to think of the origin of the Big Bang, as I discuss briefly in Chapter 14). If normalizable probability distributions, hence Possibles, are real and *are* located in both all of space and time up to special relativity, we again have an account of nonlocality. And we have an account for why measurement can instantaneously alter wave functions as when one of n entangled particles is measured, creating a new Actual where an Actual is defined as obeying the law of the excluded middle, and the wave function for the remaining $n - 1$ entangled particles instantaneously changes. Furthermore, we have an account of "which-way" information destroying interference patterns, and why measurement by looking in box 1 and *not* finding the electron in box 1, instantaneously measures the electron to now exist in box 2 despite the fact that no observation of box 2 occurred, a null measurement which can, counterfactually, have actual consequences (see Penrose et al. 1997).

I will build later on this dualism, res potentia and res extensa, to propose that conscious observation by quantum variables and us, and free-willed choice, that is, *Mind*, forms a new Triad: Actuals, Possibles, and Mind, where new Actuals acausally enable new quantum Possibles, which upon acausal measurement by *Mind* yield new Actuals yielding acausally, in turn, new quantum Possibles for Mind to measure. This will yield a highly participatory universe in which Mind, Possibles, and Matter, that is, Actuals, whether single quantum wave functions or stable "classical" spots on a silver halide screen, are at the base of the universe. This is a forever becoming, a status nascendi. This ignores the four forces, but perhaps if measurement involves fermions exchanging force-carrying bosons, the strong, weak, and electromagnetic forces are also involved. I end noting we are not certain how the classical world arises, for example on Stapp (2011), decoherence (Zurek 1991), or my try later using the quantum Zeno effect, which seems testable and could establish that quantum variables do really measure one another. We need the "classical-enough" world for the spot on the silver halide screen *is* stable for years, and were the world not classical enough, biological evolution and the accumulation of adaptations could not have occurred, organisms could not be embodied, nor could we account for the vast non-random array of fossils in the geological record. That record, unlike stars, is hugely diverse, and related in the higher taxa from phyla down to species. Any account of the classical world and its relation to quantum mechanics must allow an account of the fossil record.

If the classical world is free of Possibles, which are real in res potentia and are not in general relativity, this could suggest we may not be able, nor even *need*, to unite general relativity, having purely Actuals in a single classical space-time and where time is a dimension and does not flow, with quantum mechanics, extended successfully to special relativity where time flows. Given the long efforts to quantize gravity, as in string theory and loop quantum gravity, this suggestion will be anathema to almost all physicists. Yet the proposal of ontologically real Possibles does seem to explain the four mysteries in quantum mechanics discussed earlier, including nonlocality and thus supports the reality of potentia.

8

Beyond the Stalemate

ANSWERING DESCARTES WITH THE POISED REALM

THE PURPOSE OF this chapter is to definitively break the causal closure of classical physics by turning to aspects of quantum mechanics that allow a "quantum mind" to have repeated *acausal consequences* for the "classical" meat of the brain and body. This will be due in part to a new "state of matter," the Poised Realm, the main subject of this chapter and, I hope, an answer to Descartes.

The stalemate is simple: If we take classical physics to be causally closed, as in Aristotle's "no first cause," or uncaused event, then classical physics claims to capture *all* the causes in the rolling billiard balls given initial and boundary conditions, so the current state of the balls, or the classical physics neurons, are entirely sufficient for the next state of the classical physics balls or neurons, so there is "nothing for mind to do, and no way for mind to do it." We are left with an epiphenomenal mind at best that can witness the world but not change it. That is the conundrum left by the failure of Descartes' substance dualism, res cogitans and res extensa, that Spinoza sought to deal with by his monism, causal mechanical in one aspect, mental in the other, with no causal connection between them. Spinoza used the seventeenth-century conception of matter. Quantum mechanics and the Poised Realm take us far beyond that conception.

The Poised Realm

I now describe the recent discovery by Gabor Vattay, Samuli Niiranen, and myself of a new "realm," or state of matter, hovering *reversibly* between quantum coherence and the "classical" world. In the Poised Realm (Kauffman et al. 2014), there are two acausal means by which quantum mind can have repeated consequences for the "classical" brain and body: (1) quantum measurement, if real and acausal; and (2) acausal quantum decoherence and recoherence in open quantum systems, as described more fully later. Because these consequences are "acausal," they do not face the stalemate of classical physics and can hope to answer Descartes.

The Poised Realm now seems real, with theory and experimental confirmation described shortly. Our patent has now issued on aspects of the Poised Realm (Kauffman et al. 2014).

I begin by telling you about quantum decoherence (Zurek 1991). In a "closed" quantum system, the Schrodinger wave propagates time reversibly. At each point in time and space, for each wave, an action variable keeps track of the "phase" of the oscillation. In a water analogy, with a regular train of propagating parallel water waves, the phase information at each point is the "amplitude and phase" of a given wave at each point in time and space. In quantum mechanics, the "wave" is plotted in the complex plane as a rough circle. The amplitude of the wave is the radius from the origin of the complex plane to any point on the circle, and it is the amplitude, which is squared by the Born rule to give the probability that that wave among perhaps many will be the one that becomes "real" upon quantum measurement. The phase at any point in time and space is shown by an arrow from the origin to a specific point on the circle in the complex plane. The arrow "rotates" around the circle as phase varies.

In a closed quantum coherent system, no phase information is ever lost. Furthermore, the sum of the squares of all the amplitudes of all the waves propagating is, remarkably, always 1.0. Thus, the squared amplitude for each wave is a real probability, varying as it should between 0.0 and 1.0. Because the sum of the squared amplitudes for all the waves is always 1.0 as the Schrodinger waves propagate, physicists say that the Schrodinger waves propagate "unitarily" and hence time reversibly.

Decoherence

In the past few decades, physicists have mounted an intensive program to examine "open quantum systems" that can *acausally lose phase information to the quantum environment* (Zurek 1991). One interpretation of decoherence is that the quantum variables of the quantum environment "measure" the quantum variables of the open quantum system. If so, quantum variables do measure one another as I shall propose below. Recall that quantum theory accounts for the mysterious interference pattern of light and dark bands in the two-slit experiment by the superposition of waves from the two slits. At points on the film emulsion where the crests or valleys of two waves propagating from the two slits meet, and by positive interference, a higher crest or lower valley is found, and a measurement event occurs in which a photon is absorbed at that spot on the film emulsion, leaving a stable spot on the film after its development. But where the crest of one wave, say from the left slit, meets the valley of a wave coming from the right slit, the two amplitudes cancel out, or interfere negatively, with one another, and no photon is absorbed at that spot on the film emulsion, leading to the dark bands in the two-slit interference pattern.

Note for future discussion that we *never see* the quantum coherent "waves"; we only see the spots, the results of measurement. I will use this later to suggest that conscious experience is associated with quantum measurement, never with quantum coherent behavior. I will also show that this hypothesis is testable both in that consciousness is associated with measurement and conversely, as part of the quantum enigma, that measurement is associated with consciousness.

Decoherence is firmly established experimentally. In the two-slit experiment, as decoherence progresses, phase information is lost to the environment. In effect, the quantum system "knows less and less" where the peaks and valleys of the waves are, as decoherence progresses. The result experimentally is that the interference pattern gradually disappears and one is left with two bright spots concentrated on the film emulsion behind the two open slits. Physicists say that the "classical world" is approached arbitrarily closely, or is classical "for all practical purposes, or FAPP."

Decoherence is established in astonishing experiments. Buckminster Fullerenes are molecules with 60 carbon atoms and a beam of them can exhibit an interference pattern. Experiments for open beam systems show that decoherence occurs.

New Physics Arises in the Presence of Decoherence

The Schrodinger waves propagate time reversibly in closed quantum systems that do not lose phase information. However, in the presence of decoherence, the Schrodinger equation no longer propagates time reversibly. Decoherence is a "dissipative" term, removing the time reversibility of closed quantum systems. The results are becoming partially understood. Among these is an established failure of a standard feature of quantum systems seen in, for example, radioactive decay. In normal quantum systems, with millions of radioactive nuclei, decay occurs in any nucleus at a random time, which is Poisson distributed in time. If this is integrated, the fraction of radioactive nuclei remaining falls by a constant fraction for any fixed time duration. In particular, there is a well-defined half-life of the radioactive nuclei when half of the initial number of radioactive nuclei has decayed. At each successive half-life, a remaining half decays, yielding an exponential decay in the number of radioactive nuclei.

This result is general—the jump between two quantum states for a closed quantum system is Poisson in time—and utterly random as when it occurs and for which of many quantum variables in the same initial energy state.

In the presence of decoherence, the jump between two quantum states is *no longer Poisson* in time, with an exponentially distributed decay and half-life. This has been confirmed with cold sodium ions. New physics arises in the presence of decoherence.

Further, in the presence of decoherence, one can no longer use standard quantum mechanics to calculate the time propagation of the Schrodinger wave. Instead, there are approximate density functional methods that can be used. Presumably the time evolution of a partially decoherent system depends upon the extent of decoherence. In the Poised Realm, the degree of decoherence can increase or decrease.

The Poised Realm

The Poised Realm is depicted with two axes, a y axis and an x axis emanating from an origin as in any Cartesian coordinate system. The y axis runs from quantum coherent at the origin, upward toward the classical for all practical purposes, FAPP world, via *decoherence*. But in addition in the Poised Realm, the system can run back *down* the y axis by *recoherence* to quantum coherent behavior. Decoherence is well established. Recoherence is less well established, but assured by a theorem due to Peter Shor (1997) if information is supplied to a decohering variable, and now by experimental evidence that is reasonably convincing.

The x axis runs from "Order to Criticality to Chaos." I need to explain. In classical physics, the behavior of a system is described by its Hamiltonian. Let's start with a frictionless pendulum. As we all know, if one starts such a pendulum at any initial displacement from pointing straight down, it will oscillate forever in exactly the same periodic motion. Plot position on an x axis and velocity on the y axis, where the two axes cross at an origin, which, on the *x* axis, depicts the position where the pendulum points straight down, and minus and plus positions on the *x* axis denote its deviation to the left and right of pointing straight down. Plot velocity on the *y* axis. Velocity is a maximum in the left or in the right direction when the pendulum is pointing straight down. The direction of the velocity, which is a vector with direction, depends upon whether the pendulum is, at any instant, to the left or right of pointing straight down. Velocity is zero when the pendulum is at its extreme left or right deviation from pointing straight down. The result of plotting the time behavior of the pendulum in this *x,y* coordinate system is effectively a circle. But we all know that if we start the frictionless pendulum at different displacements from pointing straight down, it will, from each initial displacement, oscillate forever, reaching that displacement to the left or right during each oscillation. Thus, if we plot a family of these oscillations for different initial displacements, we will get a family of *concentric circles* in the *x,y* coordinate system.

In Chapter 2, I described trajectories in state spaces. Nearby trajectories can converge, stay parallel, or diverge. For example, in chaotic systems, as Poincaré showed, almost identical initial states lie on trajectories that diverge exponentially. A quantity called the "Lyapunov exponent" measures the rate of exponential divergence, where the Lyapunov exponent is positive; or convergence, where the Lyapunov exponent is negative; or where trajectories are parallel and neither converge nor diverge, where the Lyapunov exponent is zero.

For the pendulum system, the trajectories from a family of different initial displacements are parallel, and the Lyapunov exponent is zero. Now consider a new *x,y* coordinate system, with a set of ever-changing Hamiltonians on the *x* axis and the corresponding Lyapunov exponent on the *y* axis. A well-established result in classical physics is that as a set of Hamiltonians is laid out on the *x* axis in a specially ordered way, at first the Lyapunov exponent remains zero. This is the ordered regime on the *x* axis. At a special point on the *x* axis, the Lyapunov exponent *just starts* to become positive by "kinking upward" off the *x* axis. This point is called the "critical point," or critical. Further out the *x* axis the Hamiltonians correspond to ever-larger Lyapunov exponents and the classical physics systems become ever more chaotic. Thus, the region on the *x* axis beyond criticality is the chaotic regime. Please note that the curve we have drawn in this *x,y* coordinate system has a "discontinuity" or sudden change in its slope, the "kink," at criticality. The slope of the curve changes from flat in the ordered regime, where the Lyapunov exponent remains zero, and, as noted, "kinks" upward at criticality to start becoming positive. Thus, there is a *discontinuity in the derivative of this curve at criticality*. Such a discontinuity is familiar in physics and called a "second-order phase transition." These arise in many places, such as magnetization, and other areas with beautiful physics now well understood. (First-order phase transitions are exemplified by the ice/water transition, a sudden jump in the state of the system.)

Thus, in classical physics, the x axis is real and well described: order, criticality, chaos.

In the quantum coherent world, wave functions can be "localized" in space, or "extended" in space. A famous example is called Anderson localization. For a quantum system on a two-dimensional surface with large-enough "energy bumps" that are relatively randomly located, the wave function becomes localized around these bumps; this localization is called Anderson localization. But as the heights and positions of these energy bumps change to become smaller, the wave function can become delocalized and extend all the way across the two-dimensional system. You already know about this. If we are concerned with electrons in a two-dimensional medium, for example (or one- or three-dimensional), then localization makes the system an *insulator*. Electrons do not flow across the medium. If the wave function is delocalized, electrons do flow and the medium is a *conductor*. *There is a transition, the "metal insulator transition,"* where the wave function just starts to become delocalized. This metal insulator transition (MIT) is the quantum analogue of criticality in the classical world.

Thus, in the Poised Realm, the y axis depicts the degree of decoherence, and the system can pass up and down the y axis by decoherence and recoherence. The x axis depicts order, criticality, and chaos for the "classical" and the quantum world where the quantum world ranges from quantum coherent to classical FAPP as one goes up the y axis. Criticality is the MIT between localized and delocalized wave functions in the quantum coherent case and in the presence of any extent of decoherence, or location on the y axis. Thus, the "critical" line is roughly parallel to the y axis and partitions the Poised Realm into an ordered half, a chaotic half, and the critical line roughly parallel to the y axis. At the MIT, a power law distribution of small and large fluctuations in the delocalization of the wave function occurs. How these fluctuations vary with the degree of decoherence is probably not yet known.

The Poised Realm Is Real

I begin with the x axis. As you know, atoms and molecules absorb and emit photons at a specific set of energies. A famous example is the hydrogen spectrum of energy levels between which hydrogen "jumps" in quantum transitions. Bohr brilliantly predicted this spectrum in the early days of quantum mechanics by proposing that electrons orbited the nucleus and quantum *jumped* between orbits, with the energy given off by photons whose wavelengths were inversely proportional to the energy associated with the difference in energies of the two orbits between which the electron jumped.

Simple and complex molecules have few or many absorption/emission "lines" from low to high energies. Consider a single kind of molecule with a set of many such energy absorption/emission energy levels and write down the energy difference between each pair of adjacent absorption/emission energy levels. Collect this data and now plot a new plot. Plot on the x axis the difference in the two energy levels. On the y axis, plot the number of cases within a narrow "bin" on the x axis of energy level differences in the list. Plot small energy differences near the origin on the x axis and larger energy differences further out the x axis. One obtains a distribution of the frequencies of energy level jumps in the spectrum of any given molecule.

This distribution has three forms, corresponding to the ordered regime, to criticality, and to the chaotic regime. The formulas for these distributions are well known. In words, in the ordered regime, the distribution of energy level jumps falls off exponentially. There are many cases for such a molecule of small energy differences between absorption/emission energies, and fewer and fewer cases of large energy differences. In the chaotic regime, the distribution is unimodal, zero at the origin of the x,y plot, and rising to a maximum on the y axis, which is rather far out the x axis. For critical systems, the corresponding curve is again unimodal, but the peak, or maximum on the y axis, is closer to the origin on the x axis than the maximum on the y axis is for chaotic systems.

Vattay and colleagues (2015) have recently measured about a thousand individual organic molecules, small organic molecules, proteins, and DNA and RNA. The results are astonishing. Almost half are in the ordered regime. But almost half are at a single point: criticality! A few are in the chaotic regime on the x axis (see also, Kauffman et al. 2014). In particular, several large proteins, all looked at, are critical with multifractal wave functions at the MIT, which is criticality in quantum coherent and Poised Realm behaviors. DNA and RNA are ordered. Several polycyclic hydrocarbons are chaotic. So is testosterone. Critical molecules are neither conductors nor insulators and have unique partially localized electronic properties, poised at the edge of insulation and conduction, exhibiting the multifractal wave functions.

I make three major points: (1) The x axis of the Poised Realm is real. (2) The positions of molecules on the x axis reflect their *quantum* behavior, that is, their absorption and emission spectra. Thus, the x axis is telling us not only about the classical world, but about the quantum world, where for quantum coherent systems, order is an insulator if the quantum variable is an electron, or more generally exhibits Anderson localization. For chaotic locations on the x axis, the wave function is extended, and for an electron the material is a conductor. And at criticality, the wave function is just starting to delocalize, with the noted multifractal (Kauffman et al. 2014; Vattay et al. 2015), small and large fluctuations in space and time. (3) As noted, almost half the molecules studied are critical, a single point on the x axis. This astonishing fact may reflect natural selection for organic molecules. Or the abundance of organic molecules at criticality may be more general and be found in organic molecules located in meteorites such as the Murchison meteorite formed in space about the time the Earth was formed and containing at least 14,000 different organic molecules at the femtomole concentration.

The y *Axis of Decoherence and Recoherence Is Real*

Decoherence is well established. Peter Shor proved a theorem regarding "qubits" in quantum computers. These can each begin to decohere. Shor proved that with the injection of "information," these decohering qubits could recohere to full quantum coherence. Shor's theorem is now in full use to devise quantum error correction procedures for quantum computers.

In order to tell you about the still scant experimental evidence for *recoherence*, I must tell you first about the exploding field of quantum biology.

Until a decade ago, physicists had calculated that at a body temperature of 300K, that is, above absolute zero, decoherence would occur very rapidly and coherence would decay exponentially with a half-life of a femtosecond, or 10^{-15} seconds. All physicists agreed that important quantum effects could not happen in the warm, wet environment of a living body.

But an astonishing result changed all that. A group at U.C. Berkeley (Engel et al. 2007) measured quantum coherence in chlorophyll, the light-harvesting molecule where an electron absorbs a photon, jumps to an excited state, and migrates to a "reaction center" in the chlorophyll and drives the energy requiring synthesis of glucose. This is photosynthesis and one of the main ways energy enters organisms. To their astonishment, the electron remained quantum coherent for about 1,000 times longer than expected. Moreover, the rate of decoherence has been measured and it is *not* exponential but a very slow "power law." More precisely, if one plots the logarithm of coherence on a y axis, and the logarithm of time on the x axis, one gets a straight line sloping down to the right. A straight line in a log plot says the y variable is the x variable raised to some power, here a negative power. Power law decay of decoherence is *very* slow compared to exponential decay with halving of coherence at each half-life; the half-life is linear in time, not logarithmic in time.

Long-lived quantum coherence has now been firmly established in a number of light-harvesting molecules (Engel et al. 2007). Quantum biology is exploding with other evidence that birds may sense the Earth's magnetic field by quantum effects (Gauger et al. 2011) and that even smell is not a classical receptor ligand-mediated process but the sensing of quantum oscillations (Gane et al. 2013).

The Poised Realm may very well explain the unexpected and very slow power law decay of coherence. Vattay and colleagues (2012) have proven that the rate of decoherence in a molecule depends upon location of a molecule on the x axis. The rate of decoherence is, in fact, the expected fast exponential for molecules in the chaotic regime, or ordered regime, but a slow power law for molecules at criticality! Furthermore, Vattay et al. have used experimental data for a light-harvesting molecule with power law decoherence and calculated its Hamiltonian and from that its absorption and emission spectrum. The molecule is *critical* (Vattay et al. 2015).

Thus, the x axis is real, the rate of decoherence is exponential for chaotic and ordered molecules, and a power law for critical molecules. As noted, among several thousand organic molecules almost half are ordered, half are critical, and a few are chaotic. We think the x axis of the Poised Realm at criticality may explain long-lived quantum coherence in biology.

Now I can mention the still weak evidence for recoherence. The light-harvesting molecule is housed in a protein. Recent evidence (O'Reilly et al. 2014; Tiwari et al. 2013) suggests that sound vibrations, called photons in quantum mechanics, can induce recoherence in the decohering excited electron. This would correspond to the Shor theorem. The reasonable conclusion is that recoherence is certainly possible (Shor 1997) and is very likely to occur in the real world, including biology. Further work is required to establish this firmly.

There is an entirely separate means by which recoherence can occur: quantum measurement of a decohering quantum variable can fully restore coherence by collapsing the system to a single coherent wave function (Kauffman et al. 2014).

Criticality in the Poised Realm Is Critical

In later chapters I will show you that the brain, as a classical system, is dynamically "critical," lying between order and chaos. At criticality, small and large avalanches of neural activities propagate across the entire brain, easily seen in functional magnetic resonance imaging time studies (Chialvo 2012). The size distribution of these is a power law, slope –1.5, and hence critical. As I will discuss in detail, brains and cell genetic regulatory networks seem to be critical as classical physical systems. This may be of deep functional significance, for such systems may maximize the propagation of correlated changes, hence "information" in a noisy environment, enabling the living system to sense and respond reliably to its complex world.

In the quantum world, ensemble theories based on ensembles of quantum Hamiltonians (Kauffman et al. 2014; Vattay et al. 2015), are, again, just beginning to show that criticality, the MIT, has special properties. On the Anderson localization ordered side, wave functions are localized. With electrons the material is an insulator. On the chaotic side, wave functions are fully delocalized, and with electrons the material is a metal conductor. At the MIT, the transition, wave functions have a multifractal size distribution that is also a set of power law slopes whose average is 0.5 in a log-log plot, indicative of criticality. For molecules this suggests a unique distribution of electron transport, neither insulating nor conducting. These delocalized wave functions carry "information," as we will see when we discuss the quantum enigma, *if* it is real. Measurement converts the quantum coherent state that does not obey the law of the excluded middle to Actuals that do obey that law. So measurements of a critical MIT system can presumably have Actual consequences across many spatial scales in the system; these measurement outcomes also may be power law distributed and may be some quantum version of the functional virtues of classical criticality in the brain and cell regulatory networks. Despite Chapter 4 claiming that no laws entail the evolution of the biosphere, selection in evolution and neuroplasticity in one life may achieve a universal of profound functional significance, classical and perhaps quantum and Poised Realm criticality.

Life in the Poised Realm?

We think life is classical physics. In pharmacology we think of receptor-ligand complexes as classical locks and keys. Vattay has recent evidence concerning the conducting behavior of molecules that suggests that critical molecules, poised at the MIT, may have optimal receptor ligand effectiveness, not if classical, but if partially quantum coherent at some point in the Poised Realm! Decoherence, compared to full coherence, may localize the wave function in a way that helps speed search for the reaction center target in light-harvesting molecules. Amazingly, Vattay and Csabai (2015) have shown that there is an *optimal degree of decoherence* in the Poised Realm that is best for some biomolecular functions. Also, Vattay pointed out at a public talk at a CERN origin of life meeting in 2012 that in the Poised Realm a single molecule could simultaneously "try" many different reactions. If so, we will have to rethink reliance on classical chemistry in biology. The extent of quantum coherent and *Poised Realm* behaviors in life may be very widespread. The Poised Realm may transform our thinking in biology, mind/brain and beyond.

Answering Descartes

The causal closure of classical physics *is* the stalemate to get beyond, at most, an epiphenomenal mind. I mentioned in the introduction to this chapter that there were two means by which a quantum mind might afford *acausal* means by which quantum mind could have acausal *consequences* for the meat of the "classical" brain and body: (1) decoherence to classicality FAPP is acausal. The phase information is lost from the open quantum system acausally. (2) Quantum measurement can not only restore coherence, but, as in the two-slit experiment, yield a "classical" spot on the developed film emulsion. Now if there is *no* known mechanism for measurement and measurement is real, then measurement, if there is no causal mechanism, can acausally alter the classical world, albeit with the physicists' doubts about what the classical world "is."

Thus, acausal decoherence and measurement afford two acausal means for quantum mind to have consequences for classical brain/body/world. But were this to happen only once in a lifetime, it would hardly suffice; decoherence must be able to recur. But this can happen either if recoherence can happen, as per Shor and beginning data on photons yielding recoherence of decohering excited electrons, and if quantum measurement can induce a decohering quantum variable to be coherent again.

We appear to have broken the stalemate due to the causal closure of classical physics. If mind is both quantum Poised Realm and classical, then the quantum and Poised Realm aspects, via decoherence and recoherence and measurement, can repeatedly have acausal consequences for the classical or classical FAPP aspect of the mind-body problem. The big conclusion is that after 350 years, res cogitans, "thinking *stuff*," failed for having no way to alter the classical physics brain, nor any need to do so because the current state of that brain was entirely sufficient for the next state of the brain, thus "the stalemate." A quantum and Poised Realm mind can alter the "classical" brain by acausal consequences. We do not need Descartes' pineal gland. We may have answered Descartes and broken the stalemate after 350 years.

Trans-Turing Systems

I now describe briefly the concept of an entirely new "computing system" beyond the Turing machine. Recall that the Turing machine is a discrete state, 0,1, discrete time, T, T + 1, subset of classical physics. Quantum coherent behavior is real, the Poised Realm is almost surely real, and the classical world is whatever the classical world is. Consider then a system that is quantum coherent, Poised Realm, and classical, with one or many variables, receiving and emitting quantum, Poised Realm and classical inputs and outputs. Let's call this a trans-Turing system (TTS). Such a system may be constructible in the near future. It would have dynamical behaviors, quantum, partly decoherent, and classical, with similar inputs and outputs. By its classical variables, a TTS could alter the classical world. By its coherent and decoherent quantum behaviors, it could be "computing," but not in any Turing sense. Such a system is *not algorithmic* (Kauffman et al. 2014).

Quantum computers are either quantum qubit logic gates whose coherent simultaneous superpositions are hoped to solve problems far faster than classical Turing machines, or they are arranged to have Hamiltonians whose minima are the solution to a given problem. The quantum system is then allowed to evolve coherently and measured, with the hopes that on repeated runs and measurements, the minimum of the Hamiltonian will be found. There are grounds to think that decoherence is a means by which good minima of a classical Hamiltonian can be readily found. Furthermore, via partial recoherence and rounds of increasingly strong decoherence, a TTS may find minima very well.

But further, in principle, classical FAPP variables can recohere and a TTS may continue evolving in time in ways we do not yet understand. The same is true for presumably acausal measurements that leave a "classical" spot as on the silver halide screen which is new in the universe, unentailed, and constitutes a new "classical boundary condition" on the Schrodinger equation and yields thereby new possible quantum behaviors for which the new classical boundary condition is not causal, but an enabling constraint that the Schrodinger equation solutions must "fit." Such a system would be "computing," but surely not algorithmic. Furthermore, two or many TTS could be quantum entangled such that they might nonlocally affect one another. Thus, I will use this parallel to the hypothesis that the mind-body system is a TTS to suggest that telepathy and telekinesis are at least possible and the latter easily testable.

We always use our most complex systems to model the mind-brain system. In current neurobiology and computer science, that system is either the Turing machine or an outgrowth of the McCulloch-Pitts system of formal neurons, back-propagation, or Hopfield nets, all which derive from classical physics. We can probably make TTS soon. They will surely not be algorithmic, so mind need *not* be algorithmic as so many now insist in neurobiology, computer science, and beyond. Artificial intelligence has and will do amazing things. Computers can outperform humans in many ways. But we are not algorithmic for we regularly find new uses of screwdrivers, which cannot be done by propositional algorithms, and the TTS may demonstrate this experimentally.

I dream that we will construct TTS soon and that technologically it will be as important over the years as the Turing machine. Furthermore, the Poised Realm is likely to have implications in drug discovery, power transmission, and many other areas. *It is a new state of matter.*

Finally, life may well be lived partially or largely in the Poised Realm. If so, we will have to rethink biology widely. For example, receptor-ligand complexes may not be classical lock-and-key mechanisms as thought in classical biochemistry, or not always be such. Molecules in such complexes may bind classically, but their functional action may be in the Poised Realm. Vattay pointed out in a public lecture at CERN in February of 2012 that in the Poised Realm, any one molecule could, by superposition, simultaneously explore many reactions simultaneously! One hardly knows what to think about what the implications may be, for cellular biology, for life, space chemistry, and technology.

Conclusions

The Poised Realm is a new state of matter that appears able, for the first time in 350 years, to surpass the stalemate due to the causal closure of classical physics in which there is nothing for mind to do, nor any way for mind to do it. The acausal consequences of the Poised Realm by reversible decoherence and recoherence, and measurement, if real and there are no deductive mechanisms for measurement, allow quantum and Poised Realm mind to have repeated *acausal* consequences for the "classical causal meat of the brain." We have possibly answered Descartes! We can, in principle, hope for a mind that is not merely epiphenomenal, but can affect the world, so, at least ontologically, the tiger can use its conscious mind and will to better chase the gazelle.

We can hope to harness the Poised Realm to create nonalgorithmic trans-Turing systems that are *not* discrete state, discrete time subsets of classical physics, as is the Turing machine. Such a system is not, therefore, algorithmic. Life solves the "uses of the screwdriver problem," and so do we. We do *not* do so algorithmically, for no propositional algorithm can list the uses of a screwdriver or find new uses. We do. We are not algorithmic, as Penrose (1989, 1994) pointed out on different grounds.

In the next chapter, I will discuss the hypothesis that the mind-brain system is a form of TTS with consciousness associated with measurement. In Chapter 12, I will propose, with respect to the quantum enigma, that consciousness is necessary and sufficient for quantum measurement and propose the Triad, Actuals, Possibles, and Mind I have hinted at earlier, with consciousness and will at quantum variable and our levels. I note as preamble that nothing in this present chapter's third-person description yet tells us what consciousness is, nor what responsible free will may be. We have not yet glimpsed the subjective pole. We have only, at best, broken the stalemate. I discuss approaching the subjective pole in later chapters.

9

Toward a Quantum, Poised Realm, Classical Theory of Mind-Body

THE AIM OF this chapter is to begin to move us toward the subjective pole. I shall propose that conscious experience is testably associated with quantum measurement. Part of the reason for this hypothesis is that we do not experience quantum coherence, only the results of quantum measurement.

In the previous chapter, I tried to show that decoherence and recoherence in the Poised Realm plus quantum measurement, provided two means by which a quantum-Poised Realm mind could have *acausal* consequences for "classical" brain and body. This breaks the stalemate of causal closure in classical physics by which any consciousness could only be epiphenomenal and merely "witness" the world but not act on it to change it. This is a big step forward, for it takes us far beyond the stalemate. Such a mind need not be merely epiphenomenal.

However, the previous chapter is totally innocent of the subjective pole, which includes what we take to be conscious experience and responsible free will. I have already noted that responsible free will choice, where we could have chosen otherwise, requires that the present could have been different, which is allowed if quantum measurement is real and ontologically indeterminate. I note that "ontologically indeterminate" is an evocative phrase, for it strongly suggests that prior to measurement of an electron in a superposition of 50 percent up and 50 percent down, both outcomes are *Possible* prior to measurement. Hence, *Possibles*, as perhaps in my res potentia, res extensa linked by measurement interpretation of quantum mechanics, must enter the story in the standard claim that measurement is ontologically indeterminate. If so, possibles bear on free will and will be discussed in a later chapter. Of course, I have already noted that coherent behavior in superpositions, say of *n* entangled particles, does not obey Aristotle's law of the excluded middle, fitting Peirce's claim that Possibles, not Actuals or Probables, fail to obey the law of the excluded middle. This is one of my bases for postulating res potentia, ontologically real Possibles, long after Heisenberg may have done so (1958).

Hypothesis: Conscious Experience Requires Quantum Measurement

I am not the first to propose that conscious experience is associated with quantum measurement. Penrose, in the *Emperor's New Mind* and *Shadows of the Mind* (1989, 1994), made the same proposal years ago, as did H. Stapp (2011). In Penrose's case, he proposes a physical "mechanism" for measurement, orchestrated objective reduction (Orch OR), as I will discuss later. He then associates consciousness and choice with objective reduction. I make the perhaps simpler hypothesis that consciousness is associated with measurement, and in a later chapter, consciousness is necessary and sufficient for measurement. Unlike Penrose (1987, 1994), I do not propose an objective reduction as a mechanism for measurement, and note that OR proposes no specific mechanism for the choice made at reduction either. I shall propose that conscious observation acausally mediates measurement. Both hypotheses are testable.

I do not quite claim that measurement is sufficient for fully conscious human experience. Recent results have examined brain activity when a stimulus shifts very slightly from being perceived to not being perceived (Tse 2013). Using functional magnetic resonance imaging data, it is clear that wide-ranging higher brain areas are involved in conscious awareness. But I must suppose in what I write later that quantum measurement occurs widely in the brain. So such measurement may be necessary but not sufficient for fully conscious experience. Just as Penrose and Hameroff suppose Orch OR, where orchestration in microtubules organizes conscious experience (in Penrose 1994), it may be the case that in humans higher brain centers play a role in orchestrating conscious experience for which measurement is necessary, but not alone sufficient in humans.

Is this hypothesis testable? Yes, at least in initial steps. Fruit flies can be anesthetized with ether. It is standard genetics to leave a normal population and select on a subpopulation for flies that are anesthetized with lower and shorter doses of ether. Ultimately a selected population may be found requiring zero or very low ether doses to be anesthetized. It is now standard to sequence the genome of the selected and unselected, or "wild-type," fly populations to see which genes and other factors such as *cis*-acting regulatory sites, microRNA sequences, and functional RNA sequences encoded in what used to be called "junk" DNA, have mutations, ranging from nucleotide substitutions to deletions, fusion of two or more genes, and so on. Assume a set of such mutations is found. Since my hypothesis is that measurement is necessary for conscious experience, but may also require higher brain centers, we know those higher brain areas and can improve that knowledge. Meanwhile we can ask, in the simple case of mutants in genes encoding proteins, what those proteins are, and where in the body, including brain, they may be located. This is straightforward biology.

To take a concrete hypothesis I like: suppose that the quantum measurement events occur in synapses, either pre-cleft, cleft, or in the post-cleft neurotransmitter receptor complexes whose behaviors modify adjacent dendritic transmembrane potentials which potentials propagate in dendritic "computations," including propagating perhaps by gap junctions, and cumulate at the neural cell body axon hillock, where if above threshold, the neuron unleashes an action potential. Before I go on, note that the previous sentence is a hypothesis that directly links potential quantum measurement events with classical neurobiology. For example, measurement might testably change the conformation of a neurotransmitter receptor complex protein, altering its activity with respect to adjacent dendritic transmembrane potentials. We could then compare the "mutant" protein in the selected low-dose duration ether selected flies and the wild-type protein from the unselected fly population to see if the latter but not the former was associated both with a quantum measurement event and with conformation change in the receptor complex protein. Alternatively, the Penrose-Hammeroff suggestion that microtubules are the essential sites of quantum behaviors underlying consciousness may be right and the mutants may concern them and alter their quantum behaviors. Microtubules are widely dispersed in most cells in most organisms and are in synapses as well. Such evidence would begin to suggest that, in fact, quantum measurement might be associated with conscious experience. Here, the conscious experience of the wild-type fly and lack of conscious experience of the mutant fly yield wild-type and mutant proteins in which the wild-type protein undergoes a conformation change that alters transmembrane potentials of adjacent dendrites. But the mutant protein does not undergo that conformation change, and no alteration in dendritic transmembrane potentials occurs. Then we can test whether the mutant protein does not carry out a quantum measurement that the wild-type protein does carry out, and if the latter measurement is associated with the conformation change in the wild-type protein, then the alteration in dendritic transmembrane potential. This is consistent with conscious experience being associated quantum measurement, which in turn alters the classical world, here the conformation of the receptor protein, that then alters neural transmembrane potentials, which in turn may alter summation at the neuron axon hillock and trigger an action potential.

But we might find in addition to this that the receptor protein was *not* acting classically, but in the Poised Realm. Suppose we found that the mutant postsynaptic receptor protein did *not* operate biologically as a classical protein with a single confirmation at each instant, but as a critical on the *x* axis molecule partially in the Poised Realm, hence *simultaneously in some number of conformations*, while also propagating quantum partially decoherent Poised Realm behaviors in time. Then the entire cascade becomes partial "measurement" as proposed in decoherence (Zurek 1991) and alters the protein "conformation" but that conformation is not a single classical conformation at any instant, but a simultaneous superposition of conformations partially decoherent, by which the measuring event altered the simultaneous superposition of conformations of the receptor protein propagating partially decoherent quantum behaviors over time in the Poised Realm. In this case, the entire behavior is, to this point, not classical at all. We must now take account of a cascade of quantum measurement and Poised Realm behaviors by the receptor protein as part of its biological action, perhaps ultimately on classical transmembrane potentials in adjacent dendrites. But this "is," or would be if found, an analogue of the trans-Turing system discussed at the end of the previous chapter. The entire mind-brain system is operating in the quantum coherent plus Poised Realm as well as acting classically, and measurement is altering the Poised Realm behavior of the superposition states of the receptor protein and perhaps, as in the spot on the silver halide emulsion, altering classical behavior as well.

Were we to find such results, with the receptor protein acting in the Poised Realm, we would begin to believe that the mind-brain operated in the quantum coherent plus Poised Realm as well as classical realm, that consciousness was associated with the wild-type protein, and not with the mutant protein from flies selected to be very easily anesthetized with ether or unconscious even without any ether, hence that consciousness was associated with measurement.

Such experiments and results are not far-fetched. Some analogue of them is open to investigation. It might be that quantum measurement is necessary anywhere in the brain for consciousness to arise, even if that consciousness also requires higher brain centers. Or it might be that measurement is only required in the relevant higher brain centers. We can investigate both.

Standing the Brain on Its Head

In standard neurobiology considering classical neurons, there is considerable debate about what the neural "code" may be. The earlier history starts with the discovery by Hubel and Weisel of receptor fields in retinal ganglia neurons. Intracellular recordings showed for a single neuron that a spot of light directed at a spot on the retina could activate the neuron spike rate, but if moved slightly in any direction, would inhibit the spike rate. These were mapped and called "on center off surround receptor fields." Conversely some receptor fields were off center on surround. This work won a Nobel Prize for each. Later it was found that some neurons, for example in the visual cortex, increased firing rate, that is, action potentials, when the eye was exposed to a narrow rectangular bar, but only if the bar was oriented at a specific angle in the visual field. These are thought to play roles in detecting edges. Mapping these onto the cortex, one finds what are called columns, each responding to a bar with a specific orientation. Across the two-dimensional surface of the brain, each column can be labeled with a line showing the orientation it responds to. Wonderfully, there are many "singularities" in whose vicinity nearby columns have lines pointing in all directions.

The brain is also dynamically critical (Chialvo 2012), a point I will discuss in detail in a later chapter. In brief, one can record "avalanches" of neuron spiking across the surface of the brain. There are many small avalanches and fewer and fewer large avalanches. If one plots the logarithm of the size of an avalanche on the x axis and bins these into small bins of almost the same-sized avalanche, then plots on the y axis the logarithm of the number of avalanches in each bin, one obtains a power law distribution with a slope of -1.5. This slope is indicative of "critical dynamics" (Kauffman 1995). A large field of work on large networks of various topologies and dynamic couplings, all classical physics, show that criticality can arise, along with order and chaos. As I will discuss in a later chapter, cell genetic regulatory networks may also be critical. There are good reasons to think that such critical networks optimize reliable information transfer in the face of minor noise. This is a very large topic bearing on how cells and organisms may classify their worlds, evaluate their worlds, and act reliably in their worlds, again, discussed later.

There is also evidence for dendro-dendritic complex dynamics prior to the summation of that activity at axon hillocks to trigger action potentials propagating down axons.

A vast literature has been assembled by experimental and theoretical neurobiologists on many aspects of neural behavior, from color perception, to flicker fusion, to very many other topics.

An enormous body of work concerns synaptic plasticity, alterations in gene activity patterns, and "epigenetic changes" in modification of chromosomes with methylation that, in turn, alters gene expression and can be transmitted across a few generations.

I want to suggest that the mind-brain may be a form of trans-Turing system, with quantum coherent Poised Realm and classical behavior interwoven as sketched earlier and as testable in the modestly near future. Furthermore, I want to suggest that quantum measurement is necessary for conscious experience. This stands the business-conscious end of the brain on its head. Nonepiphenomenal consciousness is possible; mind can have consequences for classical matter. We need yet to address responsible free will in the later chapters of Part II. If something like this is true, it only *adds* to the fascinating work on the classical physical behavior of the brain coupled to the body and world via sensory and motor activities. I emphasize again that, due to the causal closure of classical physics, we face the stalemate if the mind-brain-body system is purely classical physics. A conscious mind is possible, but it can only witness the world, not alter it. There can be no coupling between that witnessing and motor activity. Such a mind is merely epiphenomenal. I hope I have explained clearly in the past few chapters that we are not limited to classical physics anymore. Quantum biology is real, even if its extent is still unknown. Given this, it is intellectually stubborn not to examine quantum coherent and Poised Realm behavior in the mind-brain system. In turn, such quantum coherent and Poised Realm behavior can have, via decoherence and recoherence plus measurement, acausal consequences for the classical aspects of the brain. Such a consciousness need not be merely epiphenomenal with no effect on the classical world. This is a large step and hope. Beyond this, we must address whether a responsible free will is possible. Consciousness and a responsible free will are the soul of the subjective pole, and this is my aim in Part II. Yet so far, all that I have said tells us, as yet, *nothing* about what consciousness *is*; it remains third-person description in the "objective pole," not yet the "subjective pole."

The Unity of Consciousness or Binding Problem and William James's Problem of Combinations

Francis Crick in *The Astonishing Hypothesis* (1995) considers the following: We see a yellow triangle and a blue square. Suppose, he says, that "yellow," "triangle," "blue," and "square" are processed in anatomically disconnected areas of the brain. I will assume that this is true, that these are processed in anatomically disconnected areas of the brain. Then why, says Crick, do we see "yellow triangles" and "blue squares," not "blue triangles" and "yellow squares"? Why is "yellow" bound with "triangle" and "blue" with "square"?

The binding problem is a subset of the problem of the unity of consciousness. Look around at your room, chairs, painted walls, edges of windows, cups, and paintings hung on the wall. You experience in your central attentive vision a "unity of consciousness," the very one that Russell hacked into sense data bits. How many features can you perceive beyond yellow and triangle? Thousands? Tens of thousands?

An older hypothesis, avoiding the unity of consciousness issue, was that there were grandmother cells that fired if and only if you saw your grandmother. Indeed, we do have neurons that respond to faces, including smiling and frowning faces. But as Crick points out, if we can discriminate thousands of features and these come in collections of many that are "bound," the number of possible combinations is vastly greater than the 100 billion neurons we have. Not all these can each be "recognized" by the appropriate grandmother-like cell.

There is a current reigning hypothesis to achieve binding. Groups of neurons in your brain oscillate at different frequencies; a common one is 40 Hertz, seen easily on electroencephalograms. The idea is that if the "percepts" of "yellow" and of "triangle" arise at the *same* phase of the oscillation, they are bound, and if "blue" and "square" arise at the *same* phase, but different from the phase for "yellow" and "triangle," they too are bound.

Well, it is possible. My skepticism is that we must bind thousands of sets of perceived features. If each requires a *different phase of the oscillation*, there must be some minimal time difference between adjacent sets to be separately bound, which is too small to be discriminated, so the sets would merge. If we want a unity of consciousness with thousands of sets, each of many percepts to be bound, how do we fit them all around the 40 Hertz oscillation such that they can be separately bound? It may work, despite my worry.

I want to propose a testable alternative for the unity of consciousness: quantum entanglement. Recall that quantum variables can become entangled; for example, two electrons can be entangled. Mathematically, this means that the wave equation for the two cannot be written as two *separate* wave equations, but as a single wave equation. In a deep sense the two entangled particles are "one." We cannot speak of separate particles and indeed while quantum coherent, we cannot speak of the state of the system at all, for it does not obey the law of the excluded middle.

A set of n entangled particles can, however, be measured in n measurements. A critical feature of this measurement is that the outcomes are correlated—the basis of nonlocality. Furthermore, the greater the number of entangled variables, the more correlated are their measurement outcomes. But then if consciousness is associated with measurement, the resulting "qualia," the philosophic word for experiences, are strongly correlated, which may yield a binding and a unity of consciousness.

For this idea to work, it must be possible to entangle the appropriate or desired set of quantum variables in the brain, say post-cleft neurotransmitter receptor protein with quantum variables. And it must be possible to *not* entangle these with the many zillions of other quantum variables. The same issue holds true for the Penrose-Hameroff microtubule hypothesis.

There may be many ways to accomplish this. I sketch one next.

The brain is at a temperature of about 300K, which corresponds to photons in the midinfrared, mild heat. Such photons can easily cross the brain. And such photons can become entangled, and entangled photons have been found in cells. Thus, it is physically conceivable that a set of quantum variables, such as receptor protein quantum variables, or microtubules, can become entangled by such 300K photons. We face the issue that entanglement can be broken as well, for example by measurement. Thus, the entanglement of a large number of "desired" quantum variables is an issue to be explored.

But how might the system "address" and entangle a desired set of quantum variables and not entangle others? There may be many ways. I propose one. Atoms and molecules have absorption/emission lines, forming their spectrum. These lines are narrow; that is, they absorb and emit only a narrow range of wavelengths per line, if the system is quantum coherent. *But* if the molecule or quantum variable is increasingly decoherent, the absorption band *widens* and absorbs and emits a greater range of wavelengths. This can be used to address "desired" sets of quantum variables. Specifically, let the range of wavelengths absorbed by coherent quantum variables around a single line be M, and the range for increasingly decoherent variables be N, with

N larger than M, and where N increases with degree of decoherence and, again, is greater than M. Then photons of wavelengths in the N range, but *not* in the M range, can entangle decoherent quantum variables and not entangle those that are coherent. Thus, in principle, "desired" sets of quantum variables can be entangled. Then the measurements of these give a specific set of correlated measurement outcomes and, by hypothesis, correlated qualia. This may constitute a unity of consciousness.

But there is more: The quantum variables in the Poised Realm decohere and recohere; thus, the sets of quantum variables that can be entangled as absorption bands widen with decoherence and narrow with recoherence change with time. Therefore, different sets of quantum variables can be entangled over time, yielding *different unities of consciousness*. It is attractive to think that this can be related to our serially shifting focus of attention.

Nothing in this need imply that a shifting focus of attention is not mediated by specific brain areas.

Many other means to achieve such selective and changing entanglement may be possible across anatomically disconnected areas of the brain; or if the brain is connected despite Crick's claim, entanglement may still play the roles I here propose.

James's Problem of Combinations

William James was perhaps our greatest psychologist. He proposed "atoms of consciousness" and then wanted combinations of these atoms to yield experiences that could be both new and whole. He never solved his problem, based on the following analogy of his atoms of consciousness with bricks. Consider a house made of bricks. The house is surely different and whole, compared to the bricks of which it is built. But, pointed out James, this newness and wholeness is only seen by an outside observer. From the point of view of the bricks, there is just a pile of bricks, each independent of the other, each an independent atom of consciousness. James never solved his combination problem (1909).

Quantum entanglement may solve the combination problem: The wave function for *n* entangled particles cannot be written as *n* separate wave functions but only as a single wave function in which we cannot even speak of the *n* particles as separate; they are "one" wave function. They are combined as a whole in this one wave function, which, upon measurement, can also yield new outcomes. With *n* entangled particles, the measurement of the first changes the density matrix used to propagate the Schrodinger equation to the remaining $n - 1$ entangled particles, thereby changing the amplitudes of the wave function, hence changing the squared Born rule probabilities of the outcome of the next measurement, which in turn changes the density matrix for the remaining $n - 2$ entangled particles and the Born rule probabilities of the next measurement. In short, if measurements, indeed with correlated outcomes, are associated with conscious experience, we may achieve both a unity of consciousness and new wholes, just as James hoped, solving his problem of "combinations." Perhaps this is part of how we find, nonalgorithmically, new uses for real screwdrivers. The same processes could play a role in Peirce's "abduction," neither deduction nor induction, but finding, or intuiting, new questions and hypotheses that are not merely the recombination of old ideas. General relativity is not a recombination of older ideas in physics. Our human minds do this all the time.

Next I will use the fact that among n entangled particles each measurement can alter the probabilities of the outcomes of the next measurement to try to defeat the standard worry that quantum measurement outcomes are random; hence, they cannot support a responsible free will.

A Comment on Previous Philosophies

Spinoza sought a single substance, God, which had both a mental and a physical aspect. Crucially, he did not allow causal interaction between the mental and physical, but presumed a harmony between them maintained by God. My "tries" in the previous chapters afford the start of a response, for on the view presented, quantum acausal consequences for the meat of the brain by Poised Realm decoherence and recoherence and by measurement obviate any need to call on *causal* interaction between mind and brain. Hence, God need not maintain a harmony in a single substance, a seventeenth-century monism, as proposed by Spinoza, with noninteracting material and mental aspects. Spinoza's view is a version of panpsychism; mind is everywhere in the universe. Penrose and Hameroff in OR (Penrose 1989, 1994), and my Triad are later versions of a panpsychism. That said, this claim is, so far, merely ontological. We have, as yet, no clue as to what consciousness is, even should it be necessarily associated with measurement.

Leibnitz sought monads of three levels, the last of two with Mind. His monads did not interact, in part to avoid causal interaction. But his world consisted of only windowless monads, so to achieve coordination he presumed a pre-established harmony, mocked later as "the best of all possible worlds." Leibnitz wants mind, but he wants to avoid causal interaction between mind and material or mind and mind. My earlier efforts with acausal consequences of decoherence and recoherence and measurement for the classical meat of the brain-body, trans-Turing systems, able also to interact classically and conceivably via nonlocal entanglement, proposes an answer to Leibnitz.

Materialism, and reductive materialism in classical physics from which mind may yet "emerge," leaves us with at best an epiphenomenal mind able to witness but unable to effect the classical world. This view is rather parallel to Spinoza, where the mental and physical aspects of his monism did not interact causally, but God correlated them. The view I present allows a nonepiphenomenal mind.

Idealism as in Berkeley (1948–1957) wishes to claim that "ideas" or "mind" is fundamental, and that matter derived from it. Already the view I here present yields a proposed answer. Mind is associated with measurement that yields Actuals obeying the law of the excluded middle, including single quantum states resulting from measurement and whatever the classical world shown by the spot on the film in the two-slit experiment may be. In my "try" at the quantum enigma in a later chapter, I shall propose the very strong, but simple, hypothesis that consciousness is both necessary and sufficient for measurement to occur, both human and animal consciousness and consciousness on the parts of fermions like electrons exchanging bosons like photons and measuring one another. As noted above, decoherence is often held to be the measurement of the quantum variables of the open quantum system by the quantum variables of the environment. If true, quantum variables do measure one another, although without invoking consciously doing so. This will lead to a start of a panpsychism that yet does not include responsible free will. In my discussion of the enigma and based in part on the strong free will theorem of Conway and Kochen (2006) in a later chapter, I will propose that we and electrons also have responsible free will in "deciding non randomly," in the case of an electron to be measured as spin-up or spin-down. In a way, given the hypothesis of res potentia for ontologically real possibles in quantum coherent behavior, and res extensa, ontologically real Actuals, linked by measurement in which measurement then takes Possibles to Actuals, I am not very far from Berkeley, but for one central issue. I claim that only measurement is associated with consciousness, not quantum coherent behavior. We never see that coherent behavior; we *infer it*. We only consciously see the results of measurement. Since quantum coherent behavior precedes consciousness on this view, and is not consciousness itself but may have information concerning what is now possible, as we will see in discussing the enigma, I, unlike Berkeley, will not say matter, that is Actuals, arise directly from conscious mind, but from mind acting acausally on Possibles to yield, via measurement due to consciousness, Actuals. From this we still have, somehow, to get to a classical enough world for there to be a non-random fossil record in the rocks.

A long tradition in European philosophy includes reference to *will* and doing, as in Nietzsche (1968) and Habermas (1971). We will, I hope, find a place for responsible free will and doing in the remaining chapters of Part II, including possible answers to the quantum enigma.

The Classical World?

To this point, the discussion has left out the central issue and mystery, so far, of the "classical" world and its relation to quantum mechanics. The Copenhagen interpretation assumes, pragmatically, that quantum mechanics does not tell us about reality itself but about what we will "experience" as the classical world, the "epistemic view," but the purely epistemic view is put in some substantial doubt by Ringbauer et al. (2015). If the epistemic view is in doubt, what might replace it? The postulate of ontologically real possibles may relieve us of the need to rely on any epistemic view of quantum mechanics, as I have discussed. But that does not yet tell us what the classical world "is." Yet Newton's laws are close approximations by which we can get to Mars, and calculate the solar system, and general relativity is a triumph confirmed to 10 to the −14 (Penrose 2004). Evolution requires a classical-enough world to accumulate adaptations. Organisms have "generalized rigidity," a concept due to Philip Anderson (personal communication), where pushing on one end pushes the other end. Organisms could not be embodied without a classical world and "generalized rigidity." There are many attempts to derive the classical world from quantum mechanics, decoherence for example (Zurek 1991). Below I will try the hypothesis that the quantum Zeno effect may yield a "classical-enough" world. This attempt, testable, would demonstrate that quantum variables do really measure one another. I note again that classical physics has only Actuals, no Possibles. If Possibles are real, somehow they seem to play no or no obvious role in classical physics. If not, and Possibles *are* real, then we cannot "unite" quantum mechanics with classical mechanics unless we find a way, perhaps decoherence (Zurek 1991), the quantum Zeno effect, or otherwise, to "get rid" of the Possibles in quantum mechanics. If Possibles are real, this may suggest that attempts to quantize general relativity by, say, loop quantum gravity or string theory, neither of which includes Possibles, cannot work.

Conclusion

The proposal of the Poised Realm with decoherence and recoherence plus measurement, both of which have acausal consequences of such a "mind" for the classical meat of the brain, breaks the stalemate with which we have struggled since Descartes and Newton. The proposals are testable and thus are science. They may prove true or false. At least they are a new conceptual framework for the issues of the stalemate, even if they as yet do not get us to a subjective pole of consciousness and responsible free will. Shifting patterns of entangled quantum variables may yield a unity of consciousness and shifting patterns of attention.

10

Toward an Ontological Basis for a Responsible Free Will

IN THIS CHAPTER I wish to explore the issues of a responsible free will. I hope to find an ontological, if not yet experiential, basis for a responsible free will. We remain short, in this chapter, of the subjective pole, but approach it in the next two chapters on the strong free will theorem of J. Conway and S. Kochen (2006), in Chapter 11, and a consideration of the quantum enigma in Chapter 12, where the enigma seems to point beyond physics and can, I think, give us a clue to a subjective pole true for us and, stunningly, perhaps the universe as a whole. If so, the world is very different than we have thought, not only because res potentia, ontologically real Possibles, may be true in a new dualism with res extensa linked by measurement, but because if *we* have responsible free will and consciousness and can sometimes act in the world to change it as it affords opportunities to do so, we will consider that electrons exchanging photons and fermions exchanging bosons more generally measure on another and are do so consciously and can act with responsible free will to change the universe as it too affords opportunities to do so. As we will see, enablement seems real and is not causal. We are too trapped by limiting ourselves to causal relations among Actuals. New Actuals enable what becomes newly possible, but they need not cause anything at all in enabling those new Possibles. Changes in possibilities, as argued, can be instantaneous, enabling new Actuals, but not causing them. The next two chapters build toward this possible view, some of which is testable at present.

The standard free will arguments are as follows: (1) Per classical physics, there is no free will, the universe is deterministic, and the present cannot have been different; also consider the issues of Actualism and not knowing initial conditions. (2) Per quantum physics, free will exists but it is random, so it is ontologically free if measurement is real and ontologically indeterminate, but it is *not* responsible. My aim is to show that quantum measurements of entangled variables can yield an ontological basis for a responsible free will.

Responsible Free Will

I first, briefly again, consider classical physics, for example Newton. Here the becoming in time of the universe is deterministic. We can have no free will, since whatever we do is determined. As I wrote earlier, Daniel Dennett argues that "could" means "I was competent to sink the golf putt." But then what does it mean to say, "I could have not sunk the putt"? Does it mean to us that I merely could have failed to sink the putt, competence failing me? Or do we mean that I could have chosen to smash the ball into the golf clubhouse, breaking the window and knocking lettuce onto someone's lap? Having asked Dennett, he kindly accused me of "Actualism," the apparently naive view that determinism means that only what actually happened could have happened. As I noted, we do use classical boundary conditions with a real sense of "could" or "would." Billiard balls rolling for a short time hit the boundary of the table only at a few spots and we calculate from Newton's laws what their dynamics will be. But in using those boundary conditions to integrate Newton's differential equations of motion and initial conditions, we also say that "were the initial conditions different, the balls would have obeyed Newton's laws and bounced in some different way." This, says Dennett, avoids Actualism. But it is a red herring. How are the "different initial conditions" to arise? Were the universe of the deistic God set in motion for Newton's laws to hold sway, only what actually happens *can* happen. Only the deistic God can have changed the initial conditions. Again as noted, we use continuous equations with Newton and with Einstein, and these use irrational numbers of a higher order of infinity than the rationals. The irrationals are infinite nonrepeating decimals. We cannot constructively evaluate most of the irrationals, although we can calculate some, such as pi. That point rightly says that we cannot specify in precise detail the initial conditions for continuous dynamical systems such as Newton's. But if we presume that the real line is real, then the initial conditions, whatever they may be as irrational numbers are what they are, so again the behavior is only among Actuals. Or we can consider an infinite set of closely related real initial conditions with an infinite set of nonrepeating decimal values beyond some truncation and calculate the diverse behaviors for these unknowable details of the initial conditions. The last move does seem to broaden Actualism to an unstatable and unknowable but infinite set of initial conditions, each with its own Actual. Does this heroic move give us our sense of free will? I doubt it, any more than deterministic chaos does, where dynamics are deterministic, but we cannot measure initial conditions to infinite accuracy and trajectories diverge so we cannot say what actually happens. However, it is true that these considerations weaken the stringency of Actualism. It seems here, however, that the indeterminacy is epistemological, not ontological, as it is in quantum mechanics if measurement is real and indeterminate.

There is a deep reason determinism, Actualism or beyond, does not fit our sense of free will. As I have said, free will seems to require that the present could have been different in a fundamental sense. This cannot happen in classical physics, given initial and boundary conditions. But this can arise in quantum mechanics if measurement is real and ontologically indeterminate, and can have, via null measurements, counterfactual actual consequences (Penrose et al. 1997). Thus, the electron prepared with Born probabilities of 50 percent spin up and 50 percent spin down could have been measured and become spin up or spin down. Thus, on most, but not all interpretations of quantum mechanics (not Bohm) the present could have been different. Nothing forces us so far to any of the interpretations of quantum mechanics. But then why not consider and take seriously the view that measurement *is* real and ontologically indeterminate, so the present really could have been different and counterfactual statements can sometimes be true of the present. At least such a view leaves the possibility of our normal sense of responsible free will open, and it is consistent with null measurements having counterfactual actual consequences (Penrose et al. 1997).

However, the hypothesis that quantum events, where measurement is real and indeterminate, as an account of responsible free will is often said to fail. A radioactive decay occurs in my brain, I lift a pan, and I kill the old lady. "Not responsible!" says my defense attorney. It was just a random quantum event! I have a free will, ontologically indeterminate, perhaps Possibles becoming one of a set of Actuals, but this becoming is entirely random in standard quantum mechanics.

The Standard Quantum Mechanics of Entangled Particles Entirely Meets the Randomness Objection to Quantum Responsible Free Will

My first step is to consider *n* independently prepared electrons, each with Born probability to be measured 50 percent spin up and 50 percent spin down. Carry out the measurements and indeed about 50 percent will be measured to be spin up and 50 percent to be spin down, with a Gaussian distribution about this expectation value for the *n* electrons.

Does substituting a *set* of *n* independently prepared electrons help the randomness issue? *No.* The set of outcomes is a set of *independent* random outcomes.

All this changes if the *n* electrons are entangled and represented by a single, nonseparable wave function. As noted in the previous chapter, it is standard quantum mechanics for entangled particles, here *n*, that whatever the outcome of the first measurement is, that outcome, say up or say down, changes the density matrix used to propagate the Schrodinger equation for coherent systems, thereby altering the amplitudes and thus the Born rule probabilities for the outcome of the next measurement among the $n - 1$ remaining entangled electrons. After that measurement, the density matrix is again changed and the amplitudes change around the remaining $n - 2$ electrons, so the probabilities of the next measurement are, in general, different than they were prior to the first and second measurements. And so on until all *n* measurements occur.

Now consider a possible real case: Among the *n* entangled particles, the amplitude for the first measurement might be 100 percent up, 0 percent down. The outcome, ontologically indeterminate is *up* with 100 percent probability. Now suppose that prior to this first measurement the present amplitudes for the other *n* particles than the one measured were less than 100 percent up and more than 0 percent down; say they hovered around 50 percent up and 50 percent down. But now the consequences of the first *up* measurement, occurring with 100 percent probability by the Born rule, can in an extreme case, alter the amplitudes such that the probability for the second electron to be measured is now 100 percent down and 0 percent up! In turn, the 100 percent outcome may alter the remaining amplitudes of the $n - 2$ entangled particles so that the probability of the next outcome is 100 percent up and 0 percent down. And so on among the *n* measurements. Thus, here, in this extreme but quantum mechanics–allowed case, each outcome of measurement is ontologically indeterminate, so *free*, but entirely *nonrandom*!

I claim that this example shows conclusively that the randomness argument against a quantum responsible free will fails. The outcome is ontologically free but entirely nonrandom.

However, this is not what we really want. We want there to be more than one possible outcome among the choices so the present could have been different. Fine, then let the entangled probabilities vary with each measurement among the remaining entangled particles, but by Born be less than

100 percent and greater than 0 percent. We will get some set of n outcomes. Three huge issues arise:

First, unlike the case of the n independently prepared electrons, the outcome of the set of entangled measurements is *not* independent. This seems a major step forward. The outcomes are *free*, but each is among say two alternatives, so the present could have been different. And the set of n measurements is not a set of n independent outcomes.

Second, I think we are confused by the term "responsible free human will." Suppose I said to you, "Does the newborn baby have free will to put its thumb into its mouth or climb the stairs?" *No.* The baby makes rather random motions and learns how to get its thumb in its mouth. Only after these capacities are learned does it make sense to say the baby may have a free will. And we do have a responsible free will, given our capacities, learned or inherited. But there is much more: We in fact make "responsible free will choices" among some typically statable set of alternatives, given the Actual "context" and what we want to do. I want to go to the store to buy food. OK, drive? Which route? Bike? Which route? Roller skate? OK, which route? Our responsible free will choices always occur in the context of our capacities, our context, and our purpose.

Suppose you have some capacities and I say to you: *Do anything at all that you want!* with no allusion to context. What do you do? It seems you may do nothing at all. Human responsible free-willed actions are often explained in terms of reasons, motives, and purposes, including finding unprestatable new uses of screwdrivers for a given purpose. Without context and capacities we have no reasons, purposes, or motives. Perhaps we invent a purpose on the spot. Indeed, "Do anything you want" seems ill formulated. It is like trying to list all the uses of a screwdriver which is not algorithmic with no specification of the proposed use or need or purpose. How do we choose what to do among the unprestatable set of "anything at all"? It seems we cannot choose, or choose responsibly, for we have no idea what "anything at all" for any possible purpose means. In the context of n entangled quantum variables with Born probabilities of outcomes less than 100 percent and more than 0 percent, those probabilities of those specific possible outcomes constitute the "context" for an ontologically, not yet experiential, responsible free will of n nonindependent measurements, each changing the context of the outcome of the next measurement. Each measurement outcome is an Actual that becomes an enabling constraint on the outcome of the next measurement of the remaining entangled particles. I wish to extend this for a lovely truth that I think is of high importance. Consider improvisational comedy. There are four of us. The first rule of improvisational comedy is that each of us must take what the person before us says, and build on it in a comedically appropriate way. Here is the example, pardon my language: "Bill, here is a fine silver platter with a steaming pile of horse crap!" You must build on it, and you may not say, "I don't want it. " You may say, "Super, where is my cookie cutter?" I want to say that my line to you is a new Actual that is not a cause but an Actual that is an enabling constraint that enables you to make a choice of your next line in a largely unprestatable Adjacent Possible that is comedically appropriate. When we have gone around several times, if we are any good, we have a funny skit. But note that none of us knew beforehand what we would co-create. Yet our actions are responsible free-willed choices. This parallels the n entangled particles and their successive measurements. Life is this metaphor, from the evolution of the biosphere to economic evolution, the evolution of art, law, comedy, and culture. We act responsibly often in an evolving set of Actuals providing contexts that do not cause, but enable opportunities and call forth an appropriate next response.

Now, back to our n entangled particles. In standard quantum mechanics, the Schrodinger equation is set equal to a *classical potential*. Unlike the billiard table of classical physics which plays a *causal* role in the dynamics of the rolling balls by Newton's third law, for every action and equal and opposite reaction, in quantum mechanics, the classical boundary conditions play *no* causal role at all! The classical boundary condition merely selects those eigen functions or other time-varying modes of the Schrodinger equation that must *fit* the classical boundary conditions! Thus, in quantum mechanics, the classical boundary conditions do not cause but *enable* the possible behaviors of the quantum system! But then the classical boundary conditions are Actuals that enable the possible behaviors of the quantum system while coherent. Also while decohering and recohering.

Then, considering again our n entangled particles in some classical potential, that potential enables the possible amplitude evolution of the n entangled particles. The classical boundary condition is an Actual that is an enabling constraint rather like the improvisational comedy gig among the four of us. And because the n particles are a single wave function and the measurement of each alters the probabilities of the outcome of the next measurement until all n are measured, the set of alternative outcomes of measurement changes with each measurement. But each measurement, if measurement is real, can yield, like the spot on the silver halide screen, a new in the universe "classical Actual," and that new Actual constitutes a new classical boundary condition, which in turn constitutes the changing "context" of the quantum system for which it is the boundary condition, hence the next measurement. The changing context given by the changing boundary conditions and the succession of actual results of measurement, each obeying the law of the excluded middle whether a single quantum state or a presumably "classical physics" spot on a film, is a new actual "context" that enables a new adjacent possible for the remaining $n - x$ entangled electrons. Note that if the spot is classical, it is unentailed and can create yet new classical boundary conditions on the remaining $n - x$ entangled particles. Thus, this new classical boundary condition alters nonrandomly the behavior of the remaining $n - x$ entangled particles. The total behavior is not random and not entailed. The n entangled particles cannot do "anything at all" but are each constrained by the Born rule after each successive measurement outcome. This may parallel the fact that we humans cannot "do anything at all"! Responsible free will action requires constraints and capacities.

I suggest that this discussion answers many of the standard concerns about quantum random events and a responsible free will. The outcomes of n entangled measurements can be ontologically free and entirely nonrandom. If outcomes are less than 100 percent up and more than 0 percent down, then the classical potential and each successive measurement yielding a new Actual constitute the changing "contexts" of the outcomes of the remaining measurements. These outcomes cannot be "anything at all," like human actions, but are constrained by the Born rule.

In conclusion, I hope I have convinced you that entangled quantum variables and classical boundary conditions, which may even change during measurement of n entangled particles, suffice for an ontological, not yet experiential, basis for a context dependent nonrandom free will. Here the context includes the classical boundary conditions and the sequence of new Actual outcomes of measurements among the n entangled particles which Actuals do not cause but enable changes in the behavior of the remaining $n - x$ coherent entangled particles.

The next chapter examines the remarkable strong free will theorem, which says: If the physicist has free will, then (1) nothing in the past of the universe determines the outcome of measurement, say of an electron spin up or spin down; (2) there is no mechanism for measurement, a claim independent of the same conclusion reached on res potentia and res extensa linked by measurement; and (3) the electron "freely and nonrandomly decides" to be measured spin up or spin down. This is the first use of an experiential term I know of in physics. "Non-random" hints of will and choice. In Chapter 12, I discuss the quantum enigma, which, if true, also demands a responsible free will. If that is accepted, and we all live our lives as if it were true at least some of the time, and the strong free will theorem holds, electrons have free will as well. As we will see, furthermore, n independent electrons, if measured, obey the Born rule. If they have free will and decide "nonrandomly," why is the Born rule obeyed? There just may be the inkling of "preference" and "intent" here.

II

The Strong Free Will Theorem

TOWARD THE SUBJECTIVE POLE

NOTHING IN THE earlier discussion involves "experience" in any way. I have at best afforded a way to achieve a "mind-body" system in which quantum and Poised Realm mind really can alter the "classical" world by acausal consequences for brain, so mind need not be merely epiphenomenal. That is a lot, if true. And in the past chapter I have attempted to meet many of the standard arguments that if mind were quantum, random quantum events due to measurement would arise, so might be ontologically indeterminate, hence "free," but would be random, so they cannot be a basis for a "responsible" free will. I hope that by considering the behavior of n entangled particles, where measurement of each alters the probabilities of the next measurement, such that the outcomes can be either completely nonrandom, or in any case, entirely *nonindependent*, forming an ontological, not yet experiential, basis for a context-dependent responsible free will in which the present could have been different, so choice could be real. More, as discussed in Chapter 13, with the addition of classical outcomes of measurement in an ongoing quantum, poised relam, classical system, where the outcomes of measurement alter the boundary conditions of a quantum system, the total behavior becomes ontologically free and nonrandom.

Since Descartes' res cogitans failed, and res extensa and Newton won with the triumph of classical physics and its causal closure, the subjective pole of experience has been distanced by the "hard sciences."

But a recent theorem by mathematicians Conway and Kochen (2006) called the strong free will theorem offers its own hope for the "experiential pole." The theorem states that if two physicists have independent free will, not assuming "responsibility," with measurements of two entangled spin 1 particles, physicist A measuring the squared spin component in three orthogonal directions of spin particle "a," and physicist B happens with free will to measure spin particle "b" in one of the three directions measured in "a," then the following: (1) Nothing in the past of the universe determines the outcome of measurements. (2) There can be no mechanism for measurement (independently supported by res potentia and res extensa linked by real measurement). (3) Then the stunning: The two entangled electrons jointly *decide* with free will and do so nonrandomly! This is the only theorem I know in physics that uses an experiential term, "decides," but it depends upon measurement being real, and do so nonrandomly.

It is important that nothing in this theorem requires the physicists to have *responsible* free will, as discussed next. So a sufficient condition for their free will is merely that quantum measurement be real and ontologically indeterminate. Thus, on most interpretations of quantum mechanics, but not Bohm or multiple worlds, the theorem holds on the ontological level. At issue beyond that is any experiential component, that is, a subjective pole.

The strong free will theorem, again, does *not* require responsible free will on the part of the physicists. The theorem assumes the two physicists can independently and freely choose to measure the two entangled spin 1 particles to be measured, particle "a" chosen to be measured by physicist A in three orthogonal directions, and "b," freely chosen to be measured independently by physicist B, in some direction "w," which just may "happen" to be one of the three directions in which spin particle "a" was measured. It is essential that the strong free will theorem does *not* require that the two physicists independently decide "responsibly." The theorem requires only that the two physicists can *independently choose with free will* to measure the pair of entangled spin 1 particles as just described. It suffices that at least sometimes they independently do measure the first particle in three orthogonal axes, and measure the second in an axis, "w," that just "happens" to coincide with one of the axes used to measure the first particle. But for this, it is entirely sufficient if for the first physicist, a single free-willed quantum event, measures particle "a" yield the outcome that the first entangled spin 1 particle is measured in "some" set of three orthogonal axes, and similarly a quantum free-willed measurement by the second physicist yields that the second entangled spin 1 particle happens to be measured in an axis "w," which may coincide with one of the axes used to measure the first particle. In these cases, the strong free will theorem holds and nothing in the past of the universe determines the outcomes of the measurements, there is no mechanism for measurement, consistent with res potentia and res extensa linked by measurement, and the entangled electrons "freely decide" *nonrandomly*.

The authors clearly believe the behavior of the spin 1 entangled particles upon measurement is not random. I quote from the theorem: "In the present state of knowledge, it is certainly beyond our capabilities to understand the connection between the free decisions of particles and humans, but the free will of neither of these is accounted for by mere randomness."

Toward Proto-Responsibility, Preference, and a Possible Experience of Intention

Assume the theorem holds. The physicists have nonrandom free will and so does the pair of entangled spin 1 particles. Then there is a further feature that, together with the theorem, suggests a proto-responsibility on the part of the freely deciding spin 1 entangled variables and, further, the possibility of an awareness of an "intention" by the particles. Consider 1,000 electrons that are identically prepared and are independent, to be spin up with 50 percent probability and spin down with 50 percent probability and all independent electrons are measured. Then, in fact, about 50 percent will be spin up and 50 percent will be spin down with a Gaussian distribution about this mean. But if the electrons are ontologically independent and each has "nonrandom free will," how can it turn out that in fact about 50 percent wind up spin up and about 50 percent wind up spin down? The electrons are *independent*! The easiest thought is that each electron has a "preference" to choose with 50 percent spin up and choose with 50 percent spin down. If so, each electron "decides" with a "preference" or "intention," and hence a "will" that is "proto-responsible." But then must the electron not "know" its preference at measurement? One is led to consider the possibility of "intention" and awareness of that intention. In short, we are led toward a subjective pole.

I find this set of suggestions interesting. This is the first theorem to use an experiential term, "decides," and does so nonrandomly. To fit n independent particles obeying the Born rule given the theorem, we are almost driven to proto-responsibility. What is our struggle? We know we know our subjective pole on a first-person basis. We are struggling to find any conceivable non-first-person grounds for experiential terms.

What of the experience of responsible will and preference on the part of the physicists? It seems we are not forced to this experiential conclusion in the theorem. However, the theorem assumes the two physicists can freely choose which experiment to do, that is, how to measure particle "a" or independently to measure particle "b." Further, if electrons have protowill and preference, and intent and experience of that intent, it is natural to think that by extension so do humans evolved from, and at base, constituted by such quantum variables. If we accept this, we have decircularized the strong free will theorem both ontologically and experientially.

Since Descartes' res cogitans, a substance dualism, lost to his res extensa via Newton, we have entirely lost the "subjective pole" in physics. We have, I think, now started to regain it. What we here discuss is a form of panpsychism, at least for "proto-responsible doing," "intention," and perhaps experience of will and intention by electrons and presumably other fermions, and, by hopeful extension, for physicists. I am going to propose in Chapter 12 in trying to address the quantum enigma that electrons *are* conscious in deciding, that is, upon measurement choice. The quantum enigma may or may not be real, for it assumes responsible free will by the physicist in choosing which of alternative experiments to do, but could have chosen otherwise, which I hope I have now almost justified, and it also asserts that human conscious observation can be at least sometimes a sufficient condition for quantum measurement. This latter hypothesis is testable, as I describe later, for retinal rods can detect single photons. Thus we can test if conscious response is associated with measurement. If confirmed, no easy task, then consciousness *is* sometimes a sufficient condition for measurement. Furthermore, Radin's experiments (2012, 2013) very tentatively suggest that distant humans focusing attention on the outcome of the two-slit experiment can alter measurement seen as alterations in the intensities of adjacent interference bands. If human consciousness can be shown to be sufficient for measurement, a big if, then we must ask what else can mediate measurement and the simplest hypothesis will be that electrons measuring one another, which is taken to be part of decoherence, are also conscious as part of that measurement. If not, and human consciousness *is* sometimes sufficient for measurement, then if electrons are not conscious during measurement, we would have to invent some other means beyond consciousness whereby measurement happens in electrons that "freely decide." In short, if the enigma is real, responsible free will is real, and human consciousness can be sufficient for measurement, that is indirect evidence that electrons are also conscious at measurement "decisions." We recover our subjective pole in a panpsychism.

Of course, this postulates a very strong form of panpsychism— measurement anywhere in the universe is always associated with a proto-responsible, free-willed doing, that is, deciding and doing, that is, being measured, so becoming, spin up or down, and with conscious experience. On this ground, consciousness is not an emergent property of biological evolution, but like mass and momentum, a feature of the universe itself, used by evolution for further reasons I mention next. But a possible further virtue of the assumption that electrons are conscious and decide responsibly at measurement is that it may just provide a solution to the quantum enigma, and to the famous infinite regress in von Neumann's formulation of quantum mechanics where there is an uncertain epistemic cut; at the near end, consciousness mediates measurement, and at the far end, the classical apparatus records so mediates measurement with no human consciousness. But we solve this indeterminate location of the epistemic cut: However the classical world may arise, it presumably remains also quantum in its fermions and bosons base. If these quantum variables are conscious and carry out measurement, the classical apparatus comprised of them can also carry out measurement via the conscious measurements of the fermions exchanging bosons. We obviate von Neumann's indefinite location of his epistemic cut.

Conclusions

I hope I have, by Chapter 10 and this Chapter 11, ontologically and experientially decircularized the strong free will theorem, which builds on the assumption that the physicists in the experiment herein have free will to choose independently how to measure the entangled particles, perfectly possible on quantum and Poised Realm mind if measurement is indeterminate and real, then the theorem holds: Nothing in the past of the universe determines the outcome of measurement, there is no mechanism for measurement, and the spin 1 entangled particles jointly *decide with free will* and do so *non*randomly. We can find independent grounds from the experiential term "decides" and nonrandom free will for the particles for a proto-responsibility, for if n independently prepared quantum variables say electrons prepared to be Born probability 50 percent up and 50 percent down are, as they always are, measured to be about 50 percent up and 50 percent down, we have to ask, if they freely decide, why do the results fit the Born rule? The simple idea is that each independently has a *preference*, or intent, to decide with probabilities 50 percent up and 50 percent down, so we have not only a free-willed deciding, but a proto-responsible deciding by the electrons or other fermions with alternative quantum states and potentially an awareness of that preference or intention, hence experience of will by the electron. In summary, given the strong free will theorem, we have some third-person grounds for experiential terms at the level, surprisingly, of electrons. And via Radin's tentative results, we have starting grounds to think human consciousness can affect or mediate measurement. By extension, we can hope for such experience of will and also consciousness in humans made of quantum variables. And by similar argument, we can hope for responsible free will among electrons and consciousness when they measure one another, even at the quantum basis of the classical world. I discuss this in the next chapter.

12

The Subjective Pole and the Quantum Enigma

MY HOPE IN this chapter is very large. I hope to find our subjective pole fully consistent with the context of science. Our difficulties include the fact that we know of the subjective pole by first-person experience and hope for third-person evidence. It seems worth noting how much of our practical real life assumes the subjective pole. I trained as an MD. Any neurological examination asks the patient about sensations upon stimuli and to respond to requests to push with arm up against his or her hand pressure, and hold toe up hard and his or her down pressure. Without belief in the general accuracy of those first-person reports, and the patient's ability to act as requested, the neurologist could not carry out his or her examination. The doctor takes the normal response or lack of it as clear evidence in the differential diagnosis. Without that data, he or she could not practice medicine. In general, most doctors ask patients about experienced symptoms, and to perform behavioral tasks on purpose, and the results are always central to diagnosis. Do we believe ourselves? I do and do not only because I trained and practiced medicine.

In seeking third-person data and a kind of conceptual coherence without "believing in other minds," which of course I and most of us do, I cannot prove all of what I shall propose, but some essential steps are in fact open to partially third-person experimental tests. It will lead us to a new view of the entire universe, one that is beyond contemporary physics in many ways, but consistent with it.

I shall base my analysis on the earlier chapters, particularly those in Part II, and the quantum enigma, where quantum mechanics seems to point beyond physics itself. The enigma may or may not be real, as I will explain. It is implicit in most interpretations of quantum mechanics.

Our Ordinary View of Our Human Agency

We think we sometimes have a responsible free will and could consciously choose to do otherwise. In doing so, we alter the world. Thus, we go to the store and buy flowers. But we might not have done so. We design a rocket, send it to Mars, and it lands on Mars, changing the mass of Mars and hence the dynamics of the entire solar system. But we might not have done so. We think we played a role in deciding, acting, and altering the universe. Suppose this commonsense view is right. Consider the rocket and Mars example. Mars and the rocket are Actuals that exist. Of course, both came into existence and we invented and built the rocket. These two and other Actuals constitute enabling constraints that do not cause but enable our free-willed choice to build and send the rocket crashing into Mars. We are *not* caused to build the rocket and send it to Mars. We are enabled; such a venture is "made possible" by the existence of Mars, perhaps the previous design of working rockets, and other actual features of the world such as engineering expertise and facilities in place. If this commonsense view is correct, including our conscious free-willed choices that could have been otherwise, then it cannot be true that physics alone suffices to account for the becoming of the universe. With the world as enabling constraints, we chose and did alter the mass of Mars and the dynamics of the solar system. But we might not have done so.

I am going to build toward a view of the universe in which the former is a general feature of the universe from quantum variables to us (Kauffman 2014).

No one has solved the quantum enigma, described in detail next. I try, but I am not a physicist. It is of importance in its own right and also because it seems to depend both upon responsible free will—at least by the physicist choosing the experiment to do but could have chosen otherwise—and a testable role for human consciousness in measurement. In turn, this depends upon measurement being real, which it is on most, but not all, interpretations of quantum mechanics.

von Neumann's Formulation of Quantum Mechanics

In 1933, John von Neumann published his masterwork on quantum mechanics. In it he postulated his famous "R" process, which somehow "collapsed" the wave function from any superposition state such that only one state was measured and the probability of that outcome was given by the Born rule, the square of that wave's amplitude.

In his effort, von Neumann confronted a deep issue. Consider a quantum system being measured by a macroscopic classical object. The measuring device, M_1, according to quantum mechanics, is made up of nuclei and electrons and other quantum variables and therefore the macroscopic measuring device will become *entangled* with the quantum system it measures and the joint quantum system plus measuring device, M_1, will *remain quantum*. But then we face an infinite regress, for a larger measuring device, M_2, measures the quantum plus entangled M_1, and M_2 now becomes entangled with the quantum and M_1 system and all remain quantum. On this debacle, there is an infinite regress of ever-larger measuring devices all remaining entangled, and all remaining quantum. von Neumann proposed that consciousness itself broke the infinite regress and mediated the R process by which measurement occurred and a single one of the superpositions was measured. At that point, in my language, a single quantum state is present or it is not present, so the outcome now fits the law of the excluded middle and is an Actual. This holds whether the single quantum state remains quantum and later flowers new amplitudes quadratically in time or is a "classical" spot on a silver halide film emulsion.

But no one liked this very much. After all, human consciousness is a recent event since the Big Bang and if consciousness is only found in living things, it seems very unlikely that consciousness existed 13.7 billion years ago when measurement presumably happened. Thus, physicists have considered with von Neumann an epistemic continuum from conscious observation to measurement by small and larger "classical" objects where a "record" is left, such as a track in mica left by a cosmic ray, as the range of ways measurement can happen. This range then raises the question of where the "epistemic cut" arises such that measurement happens. I hope to answer this issue by proposing that quantum variables such as electrons and protons exchanging photons consciously measure one another. Then, whatever the classical world may be, it remains at base made of quantum variables consciously measuring one another so the mica can measure the cosmic ray as it leaves a track and record.

The Quantum Enigma

I build on the examples in the book *The Quantum Enigma* (Rosenblum and Kuttner 2006). The example, say the authors, is overly simple, but the essential physics is correct.

One electron is prepared as a superposition in two boxes. Note first that "we prepared" the electron in a superposition in the two boxes, not mentioned by Rosenblum and Kuttner, but could have done otherwise. How did "we" do that? Did we have a choice? In any case, after preparation, if we are responsible and free willed, and thus counterfactual statements can be real, we can choose to do either an experiment to look in a box to consciously see if the electron is there, yes or no, or instead, counterfactually, we can choose to do an experiment that allows us to "infer" that the electron is in a superposition in two boxes. This requires that the present can have been different, which is allowed if measurement is real and ontologically indeterminate.

The first experiment is where we do see the electron in box 1. Thus, in the enigma, we, free willed, choose the question to ask of nature, one of the two experiments, then nature answers. In the case of the first experiment where we look in box 1, if we see the electron, it *is* measured by that conscious observation, and comes to exist by that measurement in box 1.

If, instead, we choose to do the second experiment that allows us to infer superposition of the electron in box 1 and simultaneously in box 2, like the two-slit experiment, we obtain the result that allows us to infer superposition. Then the enigma is that *we*, by our free-willed choice of which experiment to do in asking nature and nature's answer, which can require our conscious measurement of the electron in box 1 or inference of superposition, jointly *create* reality! Reality does not exist until our choice of experiment to ask nature a question, and nature's answer, which we measure or infer. Then we, free willed and conscious, and nature jointly co-create reality. This is the enigma. Reality seems to require *us*. Reality does not exist separately from us!

Note that we do not consciously experience the superposition of the electron simultaneously in both box 1 and 2, hence breaking the law of the excluded middle. But if we look in box 1, we *can* see the electron consciously and measurement happens. For this reason I have associated measurement as at least a necessary condition for consciousness. I will propose shortly that conscious observation is a sufficient condition for measurement, as von Neumann did. In addition, I will propose that electrons and protons exchanging photons, or fermions exchanging bosons, consciously measure one another also. This last step is essential to my discussion, requires the claim that quantum variables can "measure" one another, which is part of the account of decoherence in open quantum systems where quantum variables of the "system" are thought to be partially measured by the quantum "environment" of the open system (Zurek 1991). However, the decoherence that does propose that the quantum environment measures the quantum system open to that environment does *not* include that such measurement involves conscious observation.

Upon my proposal, conscious observation by us or by quantum variables measuring one another is a necessary and sufficient condition for measurement, part of my Triad. A virtue of this proposal is its capacity to explain much of the enigma and von Neumann's epistemic cut problem.

Another fact is essential: *if* we look in box 1 and the electron is *not* in box 1 where we look, it now "comes to be" Actual in the second box; that is, finding the electron *not* in first box "collapses the wave function" so it comes to exist in box 2, despite the fact we did not look in box 2. This is the *null* measurement we have discussed that can counterfactually have actual consequences (Penrose et al. 1997). For this to be true the quantum coherent system must somehow have information. My claim is that this "information" is nothing other than the existence of the very possibilities of the quantum coherent system that does not obey the law of the excluded middle, and if one of two possibilities, "Yes in box 1," is not true, by observation, and is now a *new Actual* that can acausally and instantaneously delete the possibility that the electron *is* in box 1. Here the fact that the electron *is not* measured in box 1 leaves only the other possibility after this new Actual, and that possibility is that the electron is measured to be in box 2, so that the single remaining possibility can and does become Actual, despite the fact that no one looked in box 2, the *null* measurement. In short, measurement happens in box 2 and the electron comes to exist as an Actual in box 2 because looking in box 1 and *not* finding the electron there removed the possibility it *was* in box 1, leaving only the possibility that it now *is* in box 2. This fits res potentia and res extensa linked by measurement. New Actuals acausally and instantaneously alter what is now possible. Note that this interpretation ignores the idea that quantum mechanics is about what we can "know," not "the world," the epistemic view favored by Bohr (1948), for which there is now contrary experimental evidence strongly suggesting that this epistemic view cannot be wholly right (Ringbauer et al. 2015).

Again, the enigma requires a responsible free-willed choice by the physicist of what experiment to do, and what he or she could have chosen otherwise; thus, it is a choice of what question to ask of nature. Free will requires that the present could have been different. If measurement is real and indeterminate, Heisenberg's potentia, or Zeh's reality of possibilities or my res potentia, and the Copenhagen interpretations all allow this. So does Whitehead (1978).

Desiderata for a Solution to the Enigma

1. What is/are the roles of consciousness in the enigma? Can human consciousness suffice for measurement? If so, might quantum variables measure one another by conscious observation so conscious observation is necessary and sufficient for measurement?

2. If we accept the strong free will theorem that the electron has free will and is nonrandomly deciding, and from the behavior of n independent electrons fitting the Born rule has preferences, hence a proto-responsibility and intent of which it may be aware, does this theorem help support our view that we too have responsible, nonrandom free will?

3. We do not know how the classical world arises. However it does, if quantum variables consciously measure and freely decide, can we understand von Neumann's epistemic range and the ambiguity about the epistemic cut?

4. If we and quantum variables such as fermions exchanging bosons are free willed and consciously measure one another, we have a panpsychism, and a participatory universe observing itself and changing what it becomes, but doing so consistent with the laws of quantum mechanics.

These points, if accepted, give us our subjective pole. Consciousness and free-willed responsible doing are present in us, in rabbits, and throughout the universe, observing and altering how it becomes as fermions exchanging bosons, consciously observe one another and, freely and responsible to the Born rule, "decide" what to do, and consciousness, perhaps acausally, mediates measurement happening.

Wheeler's It from Bit and a Participatory Universe Observing Itself

John Archibald Wheeler, who recently died, fielded three conjectural "clues": (1) it from bit; (2) a participatory universe; and (3) a universe observing itself (1990).

Here are Wheeler's words: "It from bit. Otherwise put, every 'it' ... derives its function, its meaning, its very existence entirely—even if in some contexts indirectly—from the apparatus-elicited answers to yes-or-no questions, binary choices, bits. 'It from bit' symbolized the idea that every item of the physical world has at bottom an immaterial source and explanation; that which we call reality arises in the last analysis from the posing of yes-or-no questions and the registering of equipment-evoked responses; in short, that all things in the universe are information theoretic in origin and that this is a participatory universe" (Wheeler 1990).

Wheeler often drew as a symbol a large U with an eye on the top right arm of the U, looking *at* the U. This was his metaphor for the universe observing itself. In part, this idea is that nothing is real until it is observed. Note that in the earlier quote, Wheeler uses an "apparatus" to record and measure, the far end of von Neumann's epistemic range, where the epistemic cut mediating measurement is the classical apparatus, as Bohr held.

Solving and Extending the Quantum Enigma to the Subjective Pole

I now make my points 1–4 from earlier. First, as noted, can human consciousness be sufficient for measurement, say by looking in box 1 and "seeing" the electron? We can test this in principle. A single photon can be detected by a rod in the eye. Small changes in stimuli can render that stimulus consciously observable or not—again this requires a first-person report, with large changes in observable neural activities open to third-person report. Suppose in outline we carry out such experiments in which a human or other conscious subject "sees" and is aware of a photon emitted by, say, an electron. The basic idea is that we should be able to test if conscious observation can be a sufficient condition for measurement, as von Neumann proposed for the R process. We would show by a variety of experiments that conscious observation was correlated with measurement, but that if anesthetized, not looking at the electron, not paying attention, and a variety of other stimulus manipulations altered the occurrence or not of consciousness also altered the occurrence of measurement. In short, can we test von Neumann's R process mediated by human consciousness? If the answer is shown to be *yes*, then consciousness must be real. And consciousness is a strong candidate to alone mediate measurement. Then is that mediation causal? It need not be.

We may just be getting initial evidence supporting this. Radin et al. (2012, 2013) report experiments in physics essays in which humans attending to the two-slit experiment locally or continents away appear statistically significantly to alter the interference pattern! Specifically this involved altering the relativity intensities of adjacent dark and light bands. The effects were small, .001 percent, but statistically very significant at a probability of the results by change of 1^{-8}, one in a hundred million. Since this pattern involves "spots" that constitute measurement outcomes, this is evidence, still very early, that human consciousness *can* alter the measurement outcomes, just as von Neumann proposed. The effect was not found if the "subject" was not paying attention, nor by robots. If human consciousness can "mediate" measurement, and we are convinced, then if electrons and protons exchanging photons do measure one another, either we invent another means for them to do so, or propose that they too mediate measurement as do humans, by conscious observation.

If Radin's results are confirmed and subjects continents away can affect the outcome of measurement, the natural candidate explanation is quantum nonlocality. However, the time interval between subject attention and possible alteration in the interference pattern may be too long given the spatial separation to demonstrate nonlocality. If nonlocality were confirmed, where Radin (personal communication 2015) suggests that a participant on Mars and a two slit device on Earth could demonstrate nonlocality, this would be evidence consciousness is associated with quantum mechanics, a central claim of Part II. Furthermore, this is a *testable new prediction* beyond standard quantum mechanics, for if consciousness can *nonlocally* affect measurement, the effect cannot be *causal. Then mind acausally mediates measurement.* On the Triad, Mind acausally measures Possibles, creating new Actuals that acausally yield new Possibles for Mind to measure acausally.

I now propose that electrons or protons or electrons and protons exchanging photons measure one another, as commonly held, as in decoherence (Zurek 1991), but they do so by being conscious of one another via the exchange of photons or virtual photons. In general, fermions exchanging bosons or virtual bosons measure one another, and also thereby transmit forces between them, and the measurement is mediated by conscious awareness. (I thank M. Haley for the point about bosons and forces.) Then, whatever the classical world may be, it is at base quantum, so by this quantum base, a classical device can record and measure, hopefully solving the location of the epistemic cut for von Neumann.

On this proposal, conscious awareness is necessary and sufficient for measurement anywhere in the universe. And that measurement is hopefully acausal, particularly given the strong free will theorem that denies any mechanism at all for measurement. However, Penrose proposes objective reduction (Penrose 1989, 1994), with consciousness and choice associated with objective reduction by gravitational self-collapse of superpositions of different possible space-time structures. However, as noted, he proposes no mechanism for the choice of which space-time is chosen. My hypothesis is simpler.

I do stress that my interpretation of quantum mechanics proposes onto-logically real Possibles. These, in turn, hope to explain nonlocality, instan-taneous changes in wave functions upon measurement, "which-way" information destroying interference, and the enigma of the coming into exis-tence of the electron in box 2 when box 2 was not observed but box 1 was observed and found not to have the electron when measurement yields new Actuals that acausally and instantaneously enable new Possibles. Again, I am not the first. Heisenberg (1958) proposed the same idea long ago, supported by Zeh (2007), Shimony (Penrose et al. 1997) and Whitehead is similar (1978).

The proposed sufficiency of consciousness for measurement among quan-tum variables, if true, entirely changes the world: one of many quantum waves is the one measured and becomes an Actual by obeying the law of the excluded middle. And the outcome is a free-willed responsible doing. The universe is observing what is happening and also able to "act" nonrandomly and perhaps with intent to change or choose what happens. If measurement involves fermions exchanging bosons thus with the electromagnetic, weak and strong forces, the free will of the outcome of measurement proposed by the strong free will theorem (Conway and Kochen 2006) might be accompa-nied by forces.

Both on Penrose and Hameroff's orchestrated objective reduction (Orch OR; Penrose 1989, 1994) and my try at a fuller interpretation of quantum mechanics uniting Actuals, Possibles, and Mind, the new *Triad*, measure-ment can be associated both with flashes of consciousness and with choice.

The Quantum Enigma at the Human Level

Ask a physicist if he or she can choose to look in box 1 but could have chosen to do the other experiment to establish interference instead, so the present could, counterfactually, have been different. My bet is that the physicist will typically say "yes," rather reflexively. Suppose he or she thinks *no*? Then how does he or she do physics experiments? How does he or she pick and "pre-pare" the initial conditions of the billiard balls to test Newton's equations by deducing the consequences and testing them? How does he or she "prepare" the electron in a superposition in box 1 and box 2? All of the nomothetic view of how to do science seems to require our commonsense view that we can choose which experiment to do. If not, how do we happen to pick and carry out the experiments that are appropriate to test our theories? Research grants are just stupid if we cannot design experiments to test our hypotheses. Design of rockets to go to the moon is stupid if we cannot explore alternative designs to reach our, gasp, intended purposes. But we got to the moon and Armstrong took one step for Armstrong and a huge step for mankind. In fact, the physicist *acts* as if he or she believes he or she could have done the other experiment, so acts consistent with the quantum enigma.

I note that Libet (1985) performed experiments years ago suggesting neural activity detectable before we are aware of choice, with different time delays up to 300 milliseconds, depending upon details. Libet concluded, however, that we still have a responsible free will "*veto.*" If so, we could have vetoed or not, and the present again could have been different, ruling out classical physics, but it is still ontologically fine if measurement is real and indeterminate (see Tse 2013 for a critique). Note that evidence suggests higher brain centers are required for full human consciousness. In this sense, consciousness may be necessary and sufficient for measurement, and measurement by itself may require "orchestration" in the brain to reach human consciousness (Dehaene 2014). Penrose and Hameroff's Orch OR (Dehaene 2014) has the same idea before me of orchestration, OR, carried out in their case by proposed behaviors of microtubules. Furthermore, if panpsychism is real, life has used it, and bacteria and Stentor, single-celled organisms with no neurons, may well be conscious but require orchestration to have evolved from whatever consciousness an electron may have to that of *E. coli* to us.

Quine and Conceptual Webs: What Do We Want of a Proof?

I turn next to William Van Ormen Quine and a bit of history of philosophy: (1) Hume asked what justified "induction." Nothing, he said, except that induction had worked in the past, so that justifies induction. But Hume pointed out that this is a circular argument and so it is no justification of induction at all. (2) Popper sought to obviate Hume's worry by saying science does not use induction but somehow proposes hypotheses and then seeks to "disprove them." Popper was driven to "disproof" as the criterion of science by the fact that the universal inductive claim "All swans are white" cannot be proven, but a single black swan disproves the claim. (3) Quine, in *Two Dogmas of Empiricism* (1951), devastated Popper. Quine's lovely argument: I think the world is flat; you, a modern, think it a sphere. We propose the critical experiment: We go to the seashore to observe a tall sailboat sailing toward the horizon. If the Earth is flat, the boat and mast should dwindle to a point. If the earth is a sphere, the hull should go down over the horizon before the still-visible mast. We eat lunch, watch, and sure enough, the hull goes down over the horizon, but we can still see the mast. "Fool!" you tell me, "the Earth is a sphere." "Wait a minute," I say, "maybe the ship sank!" This means we can doubt the facts of the case. But we radio it and the ship is fine. "*Fool!*" you say. "*No, no,*" I say, "I think you have assumed light rays travel in a straight line in a gravitational field. I think the hull disappeared over the horizon before the mast because light rays *fall* in a gravitational field!" Quine's wonderful point is that no hypothesis confronts the world alone, but in a "web" of claims about the facts of the situation and a web of interwoven hypotheses about the world. Popper's "disproof" is naive. We can save any one or several hypotheses we want but at the price of less or more convoluted accounts of the world. We wind up picking a relatively consistent web that seems to work and fit what we experience at the edges, says Quine. Then Einstein blows it away with fundamental changes to that web.

Quine is right, but he focused on science and third-person science "knowing" or "scientia" by and large. That is, he limited himself to propositional knowing, largely in science (after doubting how propositions reliably refer and mean earlier in the same article). Why should we limit ourselves given his wise view? Our own conscious experiences, conscious of the world and pushing with our right but not left leg against the neurologist's hand, are surely data to ourselves. So are our thoughts, as when Descartes could not doubt that he was doubting. Then are we not stubborn to ignore first-person reports?

We need a far wider conceptual web that catches much more of what we know and live. Writing Quine broader than he might want, taking Wittgenstein's language games as human data, taking the fact that we use metaphors all the time in life and they are neither true nor false as are propositions, noting that Michelangelo's *David* is not true or false, taking the fact that on many interpretations of quantum mechanics, measurement is real and ontologically indeterminate so the present could have been different, I want to claim a broader conceptual framework for us, in which we do sometimes have responsible free will, could have chosen otherwise, and are conscious. Furthermore, we consciously choose and act in the world rather as we often think we do, and rocket to Mars.

In short, I want to extend Quine to say that our criteria for "proof" are very fuzzy indeed. A web that includes quantum real indeterminate measurement, consciousness by us, testable as sometimes sufficient for measurement, the extension of that sufficiency by us, if "established—noting Quine's issues carefully—to fermions measuring one another by exchanging bosons, plus the strong free will theorem and proto-responsibility of freely deciding electrons fitting the Born rule, starts to yield a conceptual web where we and fermions have responsible free will, are conscious, can sometimes act for reasons, motives, and purposes as we think. If not, how do we understand Shakespeare's plays with their evocative metaphors? "Out out damned spot!" Well? Third-person verified science is great, but it is not all that we need to make sense of our lives in our world and be alive in that world. Furthermore, Part I obviates nomothetic science if no law entails the becoming of the biosphere. We cannot deduce entailed consequences to test in the first place.

In summary, writing Quine larger, we can have our subjective pole in a better conceptual web that catches us, *David*, and the objective world at once. We can have our humanity back!

The Triad: Actuals, Possibles, and Mind

I close this discussion by a fuller new interpretation of quantum mechanics. Earlier I have proposed res potentia, ontologically real possibles that do not obey the law of the excluded middle, and res extensa, ontologically real Actuals, that do obey the law of the excluded middle, linked by measurement. I have now also proposed the testable idea that the conscious observation by humans can suffice for measurement and extended that, were it true, to fermions exchanging bosons and consciously measuring one another. I have noted D. Radin's recent results that distant participants can alter the outcome of the two-slit experiment, altering the ratio of the intensity of the central two dark and light interference bands slightly, .001 percent but with a confidence of 10^{-8} (Radin et al. 2012, 2013). But each spot in the two-slit experiment *is* a quantum measurement event. If Radin's results are well confirmed, and we accept the first-person statement of attending or not attending and the negative results by robots, then we may come to accept that human consciousness can alter measurement. Perhaps the effect is slight because the "classical" measuring device is quantum at base and measures at that level. If we accept human consciousness can mediate measurement, then we would have to invent some other means by which classical devices measure, or quantum variables at the base of such devices measure. Then suppose we do show human consciousness suffices for measurement; hence, it becomes plausible that quantum variables in "classical devices," and electrons and protons exchanging real or virtual photons, probably measure by being conscious and they and we act with free will. Then the simplest idea is that Mind acausally mediates measurement that converts Possibles to Actuals. Thus: Actuals, Possibles, Mind. This is a huge step, but it might be right.

Now recall the store on J Street, which when closed was a new Actual that acausally and instantaneously altered what is now possible. So, too, in biological evolution by Darwinian preadaptations where the new Actual, the swim bladder, does not cause but enables a new adjacent possible set of opportunities for the evolution of the biosphere. So, too, for technological evolution where the new Actual, the personal computer–enabled word processing in the adjacent possible of the economy, which new Actual enabled file sharing, a new Actual that enabled the Web.

Note next that measurement creates new Actuals that obey the law of the excluded middle and whether a single wave function is the result of measurement or a stable "classical" spot on the screen, the new actual does not cause, but enables new quantum possibilities. For the single wave function, it will flower quadratically in time to some set of "new" wave functions, thus new possibilities if quantum superpositions are Possibles. This flowering is not causal, for the Schrodinger equation has no energy term. So the new Actual, the single wave function, enables new flowerings of more wave functions, that is, new Possibles. For the spot on the screen, it is a new "classical" Actual in the universe that is a new "classical" boundary condition on some next quantum system and as that new boundary condition the quantum waves must fit, it does not cause, but enables new quantum behaviors. In short, the Actual results of measurements enable new quantum coherent behaviors that do not obey the law of the excluded middle so are new Possibles. New Actuals acausally enable new Possibles.

Chapter 13 emphasizes that this sequence of becomings of new Actuals and new Possibles is entirely unentailed except on Bohm, which is nonlocal in the extreme and has no account for that nonlocality except to posit it as a feature of wave functions, as Einstein, Podolsky, and Rosen (1935) complained about all of quantum mechanics.

Then the Triad at the quantum level is as follows: New Actuals enable new Adjacent Quantum Possibles, which can be acted on acausally by Mind to measure to again create New Actuals that create acausally new Possibles for Mind to again act on. This is, like the biosphere, a persistent becoming, a persistent status nascendi in a panpsychist participatory universe that "is not," but "becomes."

The Triad is new because it includes Actuals, Possibles, and Mind all woven together acausally. It seems like a new view of the nonclassical world and, if the spot on the screen is classical, aspects of the classical world without yet an answer as to how the classical world comes to be out of quantum mechanics. (I try skeptically in Chapter 14.) The Triad is, at present, very distant from general relativity, which seems to only concern Actuals and their causal relations. If the Triad—with Possibles—is true, it cannot be easily united with general relativity in any simple way if general relativity and the classical world lack Possibles. In Chapter 14, I will suggest that a classical world, or classical-enough world, may arise via the quantum Zeno effect, which may by repeated measurement, repeatedly eliminate alternative possibilities, and if so, we may not need to unite general relativity and quantum mechanics.

Note how far we are here from Descartes, Spinoza, and Leibnitz. They struggled with mind and matter, and a seventeenth-century concept of matter. The Triad is a possible answer to their struggles.

Penrose and Hameroff's Orchestrated Objective Reality

Roger Penrose in *The Emperor's New Mind* and *Shadows of the Mind* (1989, 1994), proposed, as a brilliant mathematician and physicist, not only that the human mind could not be algorithmic, but mind must depend on quantum mechanics. I have been inspired by Penrose if my path is somewhat different. Objective reduction (OR), as I understand it from the books and personal communication with Hameroff online to a discussion group, proposes that quantum superpositions are branching "possibilities" in the very structure of space-time. Note that this holds that superpositions are possibilities along with Zeh (2007), Heisenberg (1958), and myself. As these alternative possible structures of space-time spread further apart under an appropriate measure of "apart," the gravitational self-interaction between these superposed "possibilities" causes collapse to one of the two possibilities for the superposition of the structure of space-time to a single, now Actual, structure of space-time. Bursts of consciousness and choice happen at reduction. The equation is $Eg = h/t$, where Eg is the gravitational self-interaction and measurement happens at a time given by h/t, where h is Planck's constant and "t" is time "flowing." This is a large extension to general relativity, where time does not flow but is a dimension and is a deterministic space-time structure. Penrose (2004) gets to his hypothesis by allowing the speed of light, C, to go to infinity, which recovers a form of Newton where time flows, hence his $Eg = h/t$, where t is time. Furthermore, as noted previously, Penrose proposes no deductive mechanism for which of the superpositions of possible space-times is the one that becomes actual, or is "chosen." This is consistent with the strong free will theorem and res potentia and res extensa linked by measurement (if measurement is real).

The Penrose view also yields a vastly participatory universe with a pan-psychism with knowing and choosing at every measurement event. In this, it is not unlike the Triad and makes testable predictions the Triad does not make. On either Orch OR or the Triad, *entangled* sets of n quantum variables, space-time itself on OR, or on the Triad, quantum variables more generally, can give rise to potentially coordinated "qualia," because measurements of each of the n alter the probabilities of the outcome of the measurements, which qualia, of the each of the rest of the still entangled variables, hence there is the potential here, as in a hope to use entanglement of anatomically disconnected areas of the brain, for a kind of "unity" of consciousness and free-willed doing in entangled sets of variables in the universe itself. For vast sets of entangled variables these could just conceivably yield forms of "cosmic mind" participating in the becoming of the universe.

My bias in favor of the Triad is in part liking ontologically real Possibles as an answer to the fact that coherence does not obey the law of the excluded middle, and an answer to why upon measurement the wave function of one or many entangled particles can change instantaneously, and thus also how nonlocality can arise acausally when, by measurements, new Actuals arise, without trying to invent noncausal "influences" that travel faster than the speed of light. I do not yet see how Orch OR helps explain nonlocality, but perhaps it does.

The Universe Can Be Entailed by No Law on Orchestrated Objective Reduction or the Triad

We are not forced at this point to either Orch OR or my Triad. Both assume measurement is real and ontologically indeterminate, but other interpretations of quantum mechanics, including Bohm and multiple worlds, do not make these assumptions. Nor do third-person ontological claims force us to first-person subjective language, including consciousness and free-willed choice. I have adduced my own grounds, including the strong free will theorem of Conway and Kochen, for the Triad, Actuals, Possibles, and Mind.

On either Orch OR or the Triad, and perhaps other views, the universe has both knowing and free willed, even perhaps proto-responsible choosing at measurement. If so, there can be no entailing law for the becoming of the entire universe as a whole, no dream of a final theory that entails all. Free-willed choice where the present could have been different as the outcome of measurement precludes entailing law of a mindless universe. Indeed, as we all know, most of us think we are sometimes free willed, and conscious and make decisions that we might have made differently, and altered the universe, such as our decision to design and build the rocket we sent to Mars, which landed on Mars, slightly changing its mass, and altered slightly the dynamics of the solar system. If we truly did so, on purpose, with responsible free will, no law entails what we did. We then are in a very creative universe. Aristotle's telos, final cause, is throughout the universe.

Conclusion

I have now completed Part II. In outline, I hope I have shown that in classical physics we can have a conscious mind, but it cannot act on the world. It is at best epiphenomenal, fruit of the causal closure of classical physics and our stalemate since Descartes and Newton. Spinoza and Leibnitz struggled to answer this based on seventeenth-century ideas of matter and mind, Spinoza with an early panpsychism.

I have introduced ontologically real Possibles, res potentia deriving from the fact that quantum superpositions do not obey the law of the excluded middle and C. S. Peirce's comment that Actuals and Probables do obey that law but Possibles do not. The hypothesis of ontologically real Possibles, not far from Whitehead, Zeh, Shimony (Penrose et al. 1997), or Heisenberg, located inside (or outside) all of space up to special relativity but also located in time up to special relativity, via an instantaneously and acausal alteration in Possibilities when New Actuals arise, offers a tentative account for why wave functions change instantaneously upon measurement and for non-locality, with no need to invent faster than light "influences," for "which-way" information altering interference patterns and why on the enigma, the electron is measured in box 2 when it was not observed but box 1 was observed to not have the electron. The Poised Realm seems real. It affords two acausal ways a quantum and Poised Realm mind can have acausal consequences for the "classical" meat of the brain, decoherence and recoherence and measurement. This breaks the stalemate of the past 350 years since Descartes and goes beyond Spinoza and Leibnitz. Furthermore, mind-body may be such quantum, Poised Realm, classical trans-Turing systems (TTS), open to construction and via entanglement and nonlocality, coordinated nonclassical behaviors. If mind-body is a TTS, then telekinesis and telepathy are possible, with the former open to clear objective tests, including Radin. TTS are not algorithmic at all in their temporal behaviors. Mind need not, as artificial intelligence asserts, be algorithmic. I join Penrose on this. This hypothesis regarding mind and body somewhat stands the brain on its head, with quantum measurement and perhaps shifting patterns of entanglement underlying a nonepiphenomenal consciousness and shifting patterns of attention, but tying in a variety of possible ways to classical neurobiology. But if the panpsychism is real, then life can have started with some form of consciousness that must have evolved very much from electrons to *E. coli* and Stentor, a single-celled organism with no neurons, to us. Classical physics cannot yield a free will both because of determinism, and because it does not allow the present to have been different. Indeterminate but real quantum measurement does allow the present to have been different. But a standard objection to quantum mind is that it may be free, but because random, cannot be responsible. I hope I have answered this by showing that measurement of n entangled particles can be entirely nonrandom or if more than one outcome of measurement is possible, the results of n measurements are now context dependent because not n random *independent* outcomes. If we include the results of measurement as new classical boundary conditions, the total behavior of the system is indeterminate and nonrandom. By the strong free will theorem, each measurement is also free willed and not random. For n entangled particles, each measurement outcome is also an enabling constraint "context" on the changed possible outcomes of the next measurements, somewhat like improvisational comedy. This discussion and the strong free will theorem established a possible ontological and experiential basis for a responsible free will among physicists and quantum variables. The quantum enigma invites testing if human consciousness can be sufficient for measurement. If so, then the easy idea is that quantum variables measure because they too are conscious. If not, we need another mechanism for reduction such as Penrose's Orch OR (1989, 1994, 2004), which violates the strong free will theorem at the ontological level. In this last chapter, braced by a wider reading of Quine's thesis about consistent webs of hypotheses and statements of facts confronting the world together to give us our accounts of the world, our personal confidence in our sometimes conscious experience and responsible doings, as real first-person data used by doctors to practice medicine, and, for most of us, our real-life practice of the conviction that we could have chosen otherwise, I take the quantum enigma seriously and conclude it is real. My Triad: Actuals, Possibles, and Mind at quantum variable and above levels, including us, is one answer to the enigma that gives us our subjective pole in a world that "is not" but that acausally "becomes" and is real in having Actuals, both single quantum wave functions after real measurement and real "classical" spots on the silver halide screen. We do not clearly know where the classical world comes from, although decoherence is one fine try (Zurek 1991). In Chapter 14, I will dare to try using the quantum Zeno effect to obtain a classical-enough world, perhaps a desperate try.

13

The Creative and Unentailed Quantum Evolution of the Universe

THIS CHAPTER IS the delayed first of the last two chapters of Part I of this book, which seeks to demonstrate that the becoming of the biosphere, social world, and here, more vastly the very universe, is creative and unentailed. My claim rests on the assumption that quantum measurement is both real and ontologically indeterminate, either as such or via Heisenberg's potentialities, Zeh's real possibilities, or my res potentia and res extensa linked by measurement.

At this point you understand the outlines of standard quantum mechanics. There are perhaps fifteen interpretations of quantum mechanics, of which I gave a new one in earlier chapters.

Central to the Schrodinger equation is that, if coherent, it propagates with time reversibility. Or, summing over the squares of all the amplitudes, these always sum to 1.0, so the equation is said to propagate unitarily.

I shall use one interpretation of quantum mechanics, the Everett multiple-world (MW) interpretation, to state my issue. On MW we can conceive of an initial quantum state of the universe, and the Schrodinger equation propagates forever unitarily. On multiple worlds, however, each time a would-be "measurement" happens, it does *not* happen; rather, the universe splits into two branches. Using the Schrodinger cat as an example, in one branch there is a dead cat and a sad physicist, but measurement does not happen. On the other branch there is a live cat and a glad physicist, but measurement does not happen. On multiple worlds, the unitary propagation of the Schrodinger equation evolves over a multiple family of disjointed universes, branching at each would-be measurement event.

Now the evolution of the Schrodinger equation is entirely deterministic given its initial conditions, so no new "possible multiple worlds" can arise that are not already included in the indefinitely expanding ensemble of multiple worlds. On this well-recognized interpretation, all possibilities, that is, amplitudes, are entailed and no unentailed amplitudes can ever arise in this "wave function of the universe."

The Creative Unentailed Evolution of the Universe at the Quantum Level

I build my case on the assumption that measurement is both real and ontologically indeterminate. This is not the multiple-world interpretation, nor Bohm's.

I wish to thank Richard Melmon for long conversations on this topic.

Consider in this light the now well-known two-slit experiment, with its light and dark stable spots on the developed film forming an interference pattern. Each spot is surely the result of measurement, in the von Neumann and Wheeler sense that it is recorded by the film, an apparatus. The spot is a measurement result.

More and critical to the last chapter, whatever the classical world is, the spot is classical, hence is a *new Actual*; the spot is either there or is not there, true or false, and was not there before that measurement.

By my hypothesis, this new actual spot is unentailed, for measurement is real and ontologically indeterminate. Furthermore, a single stable spot is a "record" of a single unentailed measurement event.

Recall that the Schrodinger equation sets his linear wave equation equal to a classical potential, U, which acts as a boundary condition that does not cause but enables the quantum behavior which mathematically must *fit* the U boundary conditions.

But the stable spot is whatever a classical object is; hence, it is a *new in the universe and unentailed* classical "object," a new boundary condition, U_1', which is also a *new Actual*.

But if measurement is real and indeterminate, U_1' is itself also unentailed as a classical object. Thus, U_1' is also a new unentailed *Actual* in the universe, which is now a new "boundary condition," U_1, on some further Schrodinger equation; therefore, that new unentailed boundary condition which the equation must fit, does not cause, but enables new quantum behavior for which U_1' is the new unentailed boundary condition. But therefore that new enabled quantum behavior, S_1', is also a new unentailed quantum behavior in the universe! Furthermore, if quantum coherent behaviors including superpositions are Possibles, the unentailed first measurement converted a Possible to a new unentailed Actual, which as a classical boundary condition created a new set of *possible* quantum behaviors that fit the new Actual classical boundary condition.

As an example, light, S_0, hitting a film emulsion yields an unentailed new spot, U_0. But then further light, S_1, hitting the developed dark spot, U_0, on the developed film will reflect differently than if no spot were there and S_1 itself may be measured, say be another film emulsion. Thus, the next new unentailed spot, U_1, the result of the second prior measurement, is also an unentailed new boundary condition, U_2, that alters the quantum behavior of a subsequent quantum system, S_2, the light hitting the new developed dark spot on the developed film.

But if U_0 was unentailed, then the new enabled behavior of S_2 is also unentailed!

In turn, S_2 may be measured to create, for example, a new classical spot U_3, which is also unentailed and enables a further unentailed quantum set of possible behaviors, S_3.

In short, if measurement is real, and ontologically indeterminate, the succession of spots, U_0, U_1, and U_2, and corresponding quantum coherent systems, S_0, S_1, and S_2, are all unentailed after the first U and S however they may have come to be.

The conclusion is that the becoming of the universe, involving quantum coherent behavior and measurement, if measurement is real and indeterminate, and can form unentailed new boundary conditions on further quantum systems that will be measured, is unentailed.

No law entails the specific history of this becoming, a sequence of unentailed results of measurement yielding unentailed new Actuals, here new boundary conditions, that in turn enable unentailed new Possibles. In short, if measurement is real and indeterminate, and can yield new unentailed boundary conditions, the history of the universe via such measurements is an unentailed history!

But so far it is not "creative" in the sense that new unpredictable possibilities can arise. For creativity it suffices that the succession of boundary conditions, U_0, U_1, and so on, not be predictable. This can easily arise if the quantum system is chaotic. Quantum chaos can easily arise, despite the linearity of the Schrodinger equation. For example, in a classical boundary shaped like a football stadium, the quantum behavior is chaotic. So if U_0, or U_1, and so on are like a stadium as boundary conditions, the quantum system will be chaotic. But that seems to mean that we cannot calculate the amplitudes of each wave, hence cannot calculate via the Born rule, the squared amplitudes, so do not know the probabilities of measurement outcomes that yield new in the universe classical boundary conditions, U_0, U_1 ... that slightly alter the shape of the football stadium. Furthermore, since we do not know U_0 or U_1 ..., we do not know the equations for the succession of quantum systems, S_0, S_1 ... Thus, we cannot we calculate the Schrodinger waves themselves, the possible outcomes, with amplitudes squared, of measurement. We know neither S_0, S_1, and so on, nor do we know U_0, U_1, and so on. As in Chapter 3, we may know less and less as U_n increases for increasing n.

Further, even without chaos, with film emulsion in the two-slit experiment, the exact form and shape and composition of the new unentailed Actual spot is not precisely predictable. But it is a new Actual "classical" boundary condition, U_1, which the Schrodinger equation must fit, so the quantum behaviors that fit the unknown details of the spot are themselves unpredictable. As in Chapter 3, over daughter and granddaughter systems, we know less and less about the untailed boundary conditions and the quantum behaviors they enable.

Therefore, the sequence of possible behaviors and measurement outcomes is not knowable, and it is creative in the sense that we cannot predict the succession possibilities, the waves, or the succession of boundary conditions that enable those waves. This parallels the last section of Chapter 3.

Therefore, a cascading sequence of unentailed and unpredictable new possibilities can arise in the evolution of the abiotic universe. The universe can thus be creative in its quantum plus real ontologically indeterminate evolution from the Big Bang. I think this means that reductive materialism fails on most interpretations of quantum mechanics. No law entails the historical becoming of the abiotic universe at these quantum levels, if measurement is real and indeterminate and there is no deductive mechanism for measurement.

Furthermore, if we adopt res potentia and res extensa linked by measurement, perhaps mind, the sequence is new Actuals, spots, are the results of measurement that form, as boundary conditions, new Possibles, that is, quantum behaviors, which then can be measured, perhaps by free-willed Mind as in the Triad or orchestrated objective reality, in a persistent status nascendi, which is unentailed, often unpredictable or unknowable at least epistemologically, in an open becoming of the universe.

This does not yet answer what the classical world is.

14

Beyond Pythagoras

MUST WE HAVE FOUNDATIONS?

SINCE PYTHAGORAS AT least, we have dreamed of ultimate scientific, hopefully mathematical, foundations. Today in physics this dream, the dream of a final theory by S. Weinberg (1994), is the reductionist dream so central to this book. If reductionism were to achieve some final theory, such that, like Newton's laws, that theory would entail all that can become in our universe or a multiverse, how would we view such a theory? There is a standard answer already clear in Newton's laws and the classical physics universe. We can ask about anything, but not "why" the fundamental laws are what they are. Those laws simply are; that's it. In a sense noted by many, the status of foundational laws that just "are" replaces God. Both are external to the universe. Then there is the deep debate about whether such laws "govern" the universe or merely describe it. Paul Davies in *The Goldilocks Enigma* (2006) notes that many physicists secretly feel that the laws "govern."

My aim in the current chapter is to explore the possibility that the view of foundations may be wrong. I hope at least to convince you that we are not *forced* to such a view; there are possible alternatives I will discuss in this chapter. If the world has no foundations, then the world will have to be far more emergent than we have dreamed. The "world" may be far more interesting and amazing if there *are* no foundations. View what I write here with deep skepticism, at most as a set of ideas to contemplate. If true, where are we led? I do not yet know, but one answer may be that our humanity in a creative universe is in a universe more creative than we have dared to imagine.

Our foundations now are in physics. General relativity and quantum mechanics, including "the standard model," the twin pillars, nonunified, of twentieth-century physics, are the best current basis of our dreams for ultimate foundations. We all know that major efforts, such as string theory and loop quantum gravity, have been underway in the past decades. Of the four forces, electromagnetic, weak, strong, and gravity, the first two are united, while the strong force is not yet successfully united with the electroweak force and gravitation is not united. String theory may succeed in uniting all four forces, but currently has 10^{500} versions, none of which can be written down. Some physicists, for example, Leonard Susskind in *The Cosmic Landscape* (2006), posit a multiverse of 10^{500} pocket universes, each with its own version of string theory; the landscape concerns how life friendly each universe is.

The issue of "life friendly" has emerged as a central problem in physics. In the standard model of quantum mechanics, with its zoo of particles plus general relativity, there are some twenty-three "constants of nature," such as the ratio of the electron to proton mass. A fundamental feature of our universe is that life exists and, with it, human physicists to wonder at the universe and what its laws might be. But it turns out that, as Paul Davies writes in *The Goldilocks Enigma*, our universe seems like a "put-up job." The constants have to be tuned to within very narrow limits to get something like our universe, with stars, galaxies, chemistry, life, and physicists to wonder at the universe. This has come to be called the "anthropic principle." How did our universe come to have the fine-tuning of the constants? One answer is God, who tunes the constants, the strong anthropic principle. The second answer is a multiverse, each with its own set of constants, and we are just lucky: Our universe happens to have the constants tuned for stars, galaxies, chemistry, and life. We could only have come to exist in such a universe; hence, the weak anthropic principle is that, due to this, of course we find ourselves in a universe with the constants tuned to allow life. *The Cosmic Landscape* is one version of the weak anthropic principle.

The weak anthropic principle commands our attention, in part, because we believe in foundations. I will discuss a conceivable alternative in which the constants (and laws) evolve; hence, they need not be tuned.

Beyond Entailing Laws in This Universe

I begin asking about foundations by recalling Chapter 4, where I hope I have convinced you that no laws entail the becoming of the evolving biosphere. Recall that we cannot prestate the functionally relevant variables—swim bladders affording neutral buoyancy in the water column, feathers co-opted from thermoregulation to flight feathers, all the Darwinian "preadaptations" we cannot prestate. But these new functional features are part of the ever-changing phase space of the evolution of the biosphere. If we cannot prestate them, we cannot write down the functional variables of the ever-changing phase space of the evolution of the biosphere; hence, we cannot write down differential equations for that evolution for we have no idea what the relevant functional variables will be; and hence, we cannot integrate the differential equations we do not have, as Newton did, to get the entailed trajectory of the specific becoming of the biosphere.

Then the evolution of the biosphere has *no foundations* in the sense of entailing law, as dreamed of in dreams of a final theory. Yet if we accept that no laws entail the detailed becoming of the biosphere and hence it has no such foundations, we also know that the biosphere is the most complex system we know in the universe. Then very complex systems can come to exist without entailing law foundations. This must cause us to begin to question whether and, if so, when and why we need foundations.

Furthermore, and this is a big issue: If the biosphere becomes without entailing laws, then a grand view of reductionism *must fail*, for the biosphere is part of the universe, and if its becoming is *unentailed*, then no final theory can entail the entire becoming of the universe or the multiverse. We are thus invited to give up the Pythagorean dream of a final entailing theory that is just there. If not, how shall we think?

I will suggest that we can view our foundational laws and constants, not as fixed, but as evolving enabling constraints that could have evolved for reasons I suggest next.

We already have "enabling constraints" as part of a new pattern of thought: foundationless, the biosphere is the most complex system we know of in the universe, with billions or more of chemical species, millions of species, a vastly interwoven web of co-evolving becoming of organisms and the worlds they mutually co-create, unknown but rich functional integration and competition. This evolution includes the very possibilities, the ever-new Adjacent Possibles, into which that evolution becomes. Recall our new, post-Newtonian framework for the biosphere, economic life, and cultural life reflecting in part the antientropic process above the level of atoms where the universe is nonergodic: New Actuals may or may not be causes, but they can also be "enabling constraints," like the swim bladder, that enable new adjacent possible opportunities into which the biosphere evolves as does our economic and cultural life. In turn, new Actuals arise in the Adjacent Possible and create yet new Adjacent Possibles in an ongoing becoming.

Among the enabling constraints in the evolution of the biosphere are physics and chemistry. If the biosphere is not reducible to physics, as I discussed in Chapter 4, because it is entailed by no laws, then the laws of physics do not cause the biosphere, but *enable* the becoming of the biosphere! Here is a first example in which physical laws are not foundational, but enabling. Without the laws of physics and chemistry, no life, no genetic code, no energy, no work cycles, no biosphere. Thus, a new thought is that physical laws can be not only causes but also enabling constraints. I will build slowly upon this to ask if the "foundational laws" of physics might just conceivably be viewed, not as just there, as just given, but as coming into existence as enabling constraints. I note that famous physicist J. Archibald Wheeler wanted the laws of physics to emerge from the "higgledy-piggledy" (1990). If so, why did "these laws" emerge, and what might they enable in the universe?

I have a hope/hunch/intuition I want to explore. You see, the biosphere has evolved, without foundations, to become enormously complex. As it has done so, what can "next arise," that is, the Adjacent Possible of the entire biosphere, has, as a secular average, exploded! The Adjacent Possible has grown, on average, over the past 3.7 billion years, and the diversity of the biosphere has become enormous. So, too, has our global economy exploded in the past 50,000 years from perhaps 10,000 goods and production technologies to 10 billion in New York alone. How has this Adjacent Possible exploded? The answer is a "purposeless teleology," as asked for by Thomas Nagel in *Mind and Cosmos* (2012), itself an expression of the antientropic processes I spoke of in Chapter 3. The diversity of ways organisms and we make a living has exploded. With this explosion, an increase in what is possible next, "the Adjacent Possible" of Part I, has exploded and the biosphere and economy are in an important sense almost ineluctably "sucked into" the very possibilities that each creates, the biosphere without selection "acting" to achieve the new possibilities afforded, for example, by the swim bladder once it exists, or the possibilities afforded by the Turing machine leading to personal computers, word processing, file sharing, the Web and selling on the Web. Our ways of being explode into the very possibilities that the explosion of new Actuals enables. The explosion is a purposeless teleology: the greater the diversity of goods, services, and production functions in the economy, the easier it is to find new combinations of the old, and equally important, new unprestatable uses of the old goods, services, and production functions to create new complements and substitute goods, services, and production functions. The economy can explode supra-critically (Hanel et al. 2007; Kauffman 1995), as diversity begets an explosion of further diversity. It is easy to see that it was far harder to invent new goods, services, and production capacities 50,000 years ago than today. Statable and unprestatable ways to combine old things and processes to create new ones grows easier with the diversity of those things and process. The same is true for the evolving biosphere.

Then why has the biosphere and economy become diverse and complex: because it is driven by purposeless teleology, allowed by the antientropic processes discussed, and as it becomes more complex, it becomes possible for it to become yet more complex. We are used to thinking of selection acting only at the level of the individual or perhaps group selection. This is not adequate. Organisms are functional wholes that achieve functional closure, or far better, functional sufficiency in their worlds that include the abiotic environment and other organisms with which they functionally collaborate and compete. Selection acts at the level of the organism in its world. A version, unrecognized, of a form of selection may act on a far wider basis than the individual or two competing groups, but at the level of communities of organisms, or more broadly at the levels of interwoven communities of organisms that ultimately comprise the entire evolving biosphere. On average, those evolutionary steps that open up wider adjacent possible opportunities for further evolution should tend to "win" as organisms in their worlds are enabled to rush *more readily* into some of those new and "broader" adjacent opportunities. In short, the biosphere as a whole may evolve such that its Adjacent Possible grows more rapidly. So, too, our evolving economy. This is a secular trend, with extinction events of species and ways of making a living included.

I want to explore, for contemplation only for now, the possibility that the same explosion of Actuals, including laws and constants, came into existence and evolved as "good" enabling constraints to allow the emergence of an ever more complex universe with an ever more, on average, rapidly growing Adjacent Possible. I want to explore the possibility that if we can have heritable different versions of the laws and constants, that version which becomes more complex with the larger Adjacent Possible "wins," the other versions "die out," and the winning version goes on winning as it evolves such that it flows into a faster growing Adjacent Possible and can become ever more complex and thus keep winning. What is "winning"? Just outpacing other versions of the laws and constants in terms of what that version of the universe and its laws and constants can become. In a sense, I want to try a broadened

"Darwin all the way down" to consider the idea that laws and constants may emerge from the higgledy-piggledy of Wheeler (1990) and evolve to enable a highly complex universe to come to exist and, more, persistently become a yet more complex status nascendi.

This strand of ideas will require a means for laws and constants to evolve with some form of heritability, as well as why diversity and complexity "win." For the latter, it may be sufficient that, like the economy becoming more complex, it is the winning of purposeless teleology; things and processes tend to become more complex, as the economy, biosphere, and universe have (Nagel 2012). I think that such purposeless teleology may be possible. So, too, may be an evolution of the laws and constants with a heritability, such that the universe becomes more complex, and the winning version of laws and constants wins. If so, the universe can "try" different laws and constants, and that set that at any moment yields the more complex universe "wins" the generalized Darwinian race. Then, just as the biosphere has become ever more complex (with small and large extinction events to be sure; Kauffman 1995), and the economy has become more complex (with small and large Schumpeterian gales of creative destruction; Hanel et al. 2007), perhaps the same can be true of the universe.

I note that the accelerating expansion of the universe allows the maximum entropy of the universe to increase, so the actual entropy, even if growing, can lag the maximum entropy so that free energy remains available to do work.

Laws as Enabling Constraints: Syntax and Grammar

I want in the next section to explore "law-like" features of language and biology that are enabling constraints. All developed human languages have both syntax and grammar. Indeed, their formal structure has been a subject of much study. Luc Steels, who works in this area, has discussed why we have syntax and grammar. The basic point is that words are comprised of phonemes and we speakers and listeners can say and hear many words. A single word, "Tiger!," is easily understood as a warning. But how can we make sense of tens or hundreds or thousands of spoken words? Syntax and grammar are the basic answers. In English, we have subject, object, noun, verb, and article, all that we are familiar with in spoken and written English. You have only to look at the phonemes, should you read aloud the paragraph I am now writing, to see that without syntax and grammar you could make no sense whatsoever of the stream of sounds you hear. Thus, critically, syntax and grammar are enabling constraints that do not cause but enable you to make sense of the words you hear. Syntax and grammar enable language and human communication. (I leave out the vexed questions of meaning and how words or phrases may mean, as the latter Wittgenstein in his *Investigations* (1953) so powerfully discussed.)

Now consider the evolved genetic code. As we all know, DNA is comprised of four nucleotides, A, C, T, and G. These are linked by $3'–5'$ phosphodiester bonds to form the two strands, call them Watson and Crick, which coil around one another to form the famous double helix, bonded by "Watson-Crick" hydrogen bonds uniquely between A and T, and between C and G. DNA is transcribed into a cousin molecule, RNA, where U substitutes for A. Transcribed RNA in the simplest case can be transformed in a few or several chemical steps into "messenger RNA." (We now know that transcribed single-stranded RNA can play many other roles in cells than merely constituting messenger RNA.)

Messenger RNA carries the famous "genetic code." Here triplets of nucleotides on the message, say "AAA" or "ACG," are "codons" of the genetic code. The linear sequence of codons in the messenger RNA will specify, via "translation," described in more detail in a moment, the linear sequence of amino acids that will be linked together via peptide bonds to form the "primary sequence" of a protein. Before we get to the code itself, a sequence of codons "codes" for a protein, but it is part of a much longer DNA sequence. Start sites, specific DNA sequences, are part of the syntax that specifies the "start," or one end, of a gene that is to be transcribed and then translated into a specific protein. Other nucleotide sequences are "stop" syntax "symbols." So the protein is specified from the DNA between syntactic start and stop site along the DNA, the triplets between will be translated, each into a specific one of the twenty amino acids used in cells.

The translation of the code is amazing. These are the fundamental steps. Small RNA molecules called transfer RNA (tRNA) have two binding sites; the anticodon site is the base pair Watson-Crick complement of the messenger RNA codon. Thus, if the codon is CCC, the anticodon site is GGG. The second site on each rRNA is the site at which "the proper" amino acid among the twenty is bound by an encoded protein enzyme called a "synthetase" to that specific tRNA. Thus, synthetase proteins encoded by the genetic code itself self-consistently "charge" each tRNA with the proper amino acid that lead ultimately to the synthesis of the very synthetase enzymes that properly charge the transfer RNAs. Thereafter the messenger RNA, with some of the amino acid–charged tRNAs bound to its codon site, is operated on by the ribozyme to link the amino acids into a growing chain that becomes the encoded protein.

In short, the genetic code involves the self-consistent charging of the right tRNA by encoded synthetase proteins which get translated into proteins including the same synthetase proteins.

No one thinking about the origin of life imagines that life started with encoded protein synthesis. Thus, somehow, encoded protein synthesis itself evolved. Why?

A simplified view of the origin of life includes four alternatives discussed further in Chapter 16: (1) Single-stranded RNA could line up free complementary nucleotides, bind them into a complementary strand, Watson from Crick, melt apart, and repeat, to form a reproducing RNA sequences. Experiments to achieve this have failed for forty years but may yet succeed. (2) Lipids in water form lipid bilayers that close to form hollow vesicles called liposomes. These can grow and bud, or divide. Some think life started by such systems. (3) I and others think life started as one molecule, or more likely a set of molecules, perhaps small proteins called peptides or RNA, or both, which mutually catalyze one another's formation from some building blocks such as amino acids or nucleotides or both. Collectively autocatalytic sets have been made of DNA, of RNA, and of proteins. Spontaneous formation of evolved RNA sequences able to act as enzymes, called ribozymes, into collectively autocatalytic sets has been achieved. We may soon show the spontaneous formation of such collectively autocatalytic sets. (4) Protocells may consist in some replicating system inside a budding liposome. We may achieve this soon. Work suggests that such systems can evolve to at least some extent, all discussed in more detail in Chapter 16.

A living cell is, in fact, a collectively autocatalytic set. However, such a protocell would have a hard time "exploring protein sequence space," while sustaining "catalytic closure" such that the collectively autocatalytic set reproduces. Thus, the critical issue is that the evolution of encoded protein synthesis, a grammar and syntax at the molecular level, did not cause, but *enabled* cellular life on Earth to readily make novel DNA sequences hence explore the vast space of possible RNA and protein sequences. Encoded protein synthesis, via a syntax and grammar, enables cells to maintain collectively autocatalytic closure and readily explore the enormous explosion of novel proteins and all their potential functionalities, in the evolution of the biosphere. Grammar and syntax in language and in cells does not cause, but enables, very complex behaviors to emerge.

The Evolution of Recombination and Sex and the Sex Ratio as a Law-Like Enabling Constraint

We have seen that no law entails the becoming of the biosphere. Once encoded protein synthesis evolved, if not before, evolution has largely occurred by mutations of DNA sequences and by what is called recombination of DNA sequences. In the simplest case, a mutation may substitute one nucleotide, say C, for a G at one spot in a DNA sequence.

Sex and recombination occurs in bacteria and sexually reproducing organisms. Consider humans and recombination. Each of us has a set of twenty-three chromosomes we inherit from mother and a similar set from father. During formation of egg and sperm cells, each of the twenty-three pairs of maternal and paternal chromosomes line up next to one another. Occasionally, a break occurs at matched sites in the two chromosomes and the "left" end of, say, the paternal chromosome attaches, or recombines, to the right half of the maternal chromosome to form a recombinant chromosome. This is recombination. From the perspective of encoded protein synthesis, recombination can create a new gene, encoding part of the maternal protein and part of the paternal protein. Thus, recombination can create very new proteins much faster than the accumulation in one organism of a set of many mutations, which substitute nucleotides such that the mutant gene is now identical to the recombinant gene. In short, recombination allows novel genes, gene combinations, and RNA and proteins to be created very rapidly, allowing rapid exploration of protein space. In addition, recombination allows two genes, A and B, initially on the maternal and paternal chromosomes, to come to be on one chromosome, aiding more rapid evolution.

A central question in evolutionary biology is why sex evolved. The basic answer is that it is thought to speed exploration of good new proteins or good new combinations of genes strung out like beads along the chromosome.

Thus, it is already clear that sex evolved because it enabled more rapid evolution. It is an enabling constraint, not a cause, rather like grammar and splicing paragraphs together.

But for sexual organisms with two sexes, it is obviously most useful if the numbers of males and females in a population are equal or nearly so. This is almost always the case; the sex ratio of most organisms is nearly 1:1. This 1:1 ratio is a law-like behavior in sexual organisms, despite the fact that no law entails evolution.

In short, a law-like behavior, 1:1 sex ratio, emerges as part of an enabling constraint afforded by recombination that does not cause, but enables, faster evolution. Law-like behaviors can emerge from enabling constraints and those very law-like behaviors, here a 1:1 sex ratio, can sustain those enabling constraints.

Might Physical Laws and Constants Evolve as Enabling Constraints?

In our standard view the laws of nature, whether they govern or describe, are "out there" to be discovered. If discovered, they just "are"; we can ask no further questions about why these laws came to be.

But we are not forced to think this way, even if it may well be correct. It seems worth considering an alternative: Laws and constants evolve. Hence, we can ask "why" they evolve.

I begin with two questions: If the laws and constants of nature evolve, why? And if so, how might they evolve?

The pivotal new idea discussed earlier, for both the evolving biosphere and economy, is that each flows into the ever-new Adjacent Possible "opportunities" that its own evolution creates. Here new Actuals create new Adjacent Possibles where the Actuals are not only causes but *enabling constraints* for further evolution. For the biosphere as a whole and the economy as a whole, evolution may occur by purposeless teleology into the ever-wider opportunities the wider Adjacent Possibilities afford, so the biosphere and economy become more complex. Furthermore, if a community faces narrower and wider Adjacent Possibles, it will tend to evolve into the wider Adjacent Possibilities.

Why has the economy and biosphere become more diverse? One answer is the purposeless teleology, where complexity begets more complexity, that Nagel seeks in Mind and Cosmos (2012). We have seen unprestatable evolution of the functionalities of the biosphere and economy. The more diversity there is, the more ways these can be recombined and, more, new, jury rigged, unprestatable functionalities found and recombined. This may well be, indeed seems to be one realization of, purposeless teleology and may drive the growth of complexity and with it an expanding Adjacent Possible that enables the very further expansion of complexity and diversity. The biosphere and economy seem to grow such that the Adjacent Possible itself grows, sucked forward into the opportunities created. Purposeless teleology begets yet more purposeless teleology, by hook or crook, prestatable or "jury rigged," not.

Note that in a deep sense, nothing "caused" the biosphere to become more diverse, nor the economy in the past 50,000 years. Both were enabled to become more diverse, driven by a purposeless teleology made possible by the antientropic processes of the nonergodic universe.

With the examples of the evolving biosphere and economy in mind, let us ask whether the same could possibly be true of the abiotic universe that has also blossomed in complexity and diversity from the Big Bang. Might the laws and constants evolve such that the diversity of ways, things, and processes can exist and interact with one another increases, creating yet new ways, new opportunities, in which new things and processes can couple to the Actuals that already exist and thus also come into existence, creating yet new opportunities for further things and processes to come into existence? And more, might those heritable versions of the laws and constants that afford wider adjacent possibilities tend to "win" at each successive step in the evolution of the laws and constants? Here "winning" is simply that that version of the laws and constants that causes/enables the universe to become more complex with an ever-greater Adjacent Possible, explored by purposeless teleology, hence "becomes" more diverse more quickly than other versions of the laws and constants. And so it can continue to become more diverse, driven by purposeless teleology and the antientropic features of the universe. It's a form of "Darwin all the way down," where the entire biosphere has also become more diverse and complex, with variations in laws and constants and the versions that yield a universe that "outpaces" the others win. If this can be possible, we can have a foundationless emergent evolutionary process for the laws, constants, and the becoming universe they enable.

Like the evolving biosphere and economy, the universe evolves into an expanding phase space of linked things and processes abetting further diversification. The phase space "expands" because, just as discussed in Chapters 3 and 4, above the level of atoms, the universe is vastly nonergodic. Not only will it not create all possible proteins with a length of 200 amino acids, as described in Chapter 4, or molecules of CHNOPS of up to 100,000 atoms per molecule, as discussed in Chapter 3, it will not make all possible stars, galaxies, planets, space grains, rocks, mountains, geysers, stream systems, or formations of diverse mineral deposits. There is an indefinite hierarchy of increasing nonergodicity as more complex things and processes and linked things and processes arise. Again this expanding nonergodic phase space, ever more nonergodic as complexity of grains, rocks, processes, and linked things and processes as complexity increases, allow the "antientropic" increase in complexity of the evolving cosmos given the free energy afforded, in part, by the accelerating expansion of the universe. This exploration is another form of purposeless teleology: complexity begets yet more complexity along pathways where what is now Actual opens new opportunities, or possible pathways for more complexity.

Our universe with its laws and constants allowed the evolution of stars and galaxies, where we presumably will not make all possible stars which form in cold molecular clouds in galaxies where the universe will not make all possible galaxies, then simple and complex atomic nuclei, complex chemistry, and space grains in the cold molecular clouds, planets, and geysers. Its Adjacent Possible is enormous in all these ways. It is well known for some different than our values of the constants that none of this could have happened, hence the weak anthropic principle and multiverse proposals.

Having said this, can we now consider whether the laws and constants of nature might evolve? We can ask whether the laws evolve and constants evolve in such a way that the universe becomes more complex and diverse, just as does the biosphere. If we can imagine "heritable" variations in the laws and constants, then, like encoded protein synthesis, those variations that yield faster growing phase spaces will be explored faster; hence, we can begin to imagine that such "versions" win simply by, again, "outpacing" versions of the constants and laws that do not expand the phase space as rapidly, that is, do not "become" as rapidly in diversity and complexity that sets the platform for becoming yet more diverse and complex. This "outpacing" is just what Addy Pross, in *What Is Life: How Chemistry Became Biology* (2012), calls "dynamic kinetic stability." Even before self-reproduction, Pross argues, the "fastest wins," that is, dynamic kinetic stability. I suggest that the same can apply to the universe becoming ever more complex, setting the ever-new stages for becoming yet more complex. To repeat a more familiar example, the global economy 50,000 years ago had perhaps 10,000 goods and production functions but has exploded in complexity and diversity, each stage enabling the next stage, to now with at least 10 billion goods and production capacities in New York alone. The economy has exploded "supra-critically (Kauffman 1995) into the very Adjacent Possibles it creates. Perhaps the universe has as well, with no fixed entailing laws.

I would note that on this view, the laws are, in a sense similar to grammar and encoded protein synthesis, enabling constraints that enable purposeless teleology, just as encoded protein synthesis does not cause, but enables wider exploration of DNA, RNA, and protein space and the exploration of often unprestatable new functionalities by which the biosphere has become so diverse and complex.

To consider this strange idea, we need to wonder about whether variations in the laws and constants can happen simultaneously, and how they might be inherited. Lee Smolin has voiced the idea of "precedence." Here constants and aspects of laws may have precedence, as in the common law. Those values of the constants or details of the laws that are "used" more have more precedence and can eventually become "fixed." Then, let's consider laws at a quantum level. Given superpositions in quantum mechanics, which do not obey the law of the excluded middle, we can add the new idea of *simultaneously* quantum superpositions of *variations* in the constants and laws. Hence, we must consider "where" the variations occur. I will assume for now, "Locally." Heritability would include precedence, and we can imagine, for these variations in the law arising at any time, entanglement of many variables with different sets of those variations in the laws, as a means of spreading "selected" values of the constants and details of the laws across regions of space and quantum variables, either in normal quantum mechanics or quantum field theories. Quantum entangled variables from the time of the Big Bang can be spread throughout the universe. Entangled variables arising later can be far flung in the universe. Heritability across vast stretches of space and time or space-time can occur.

If we try these ideas, they are surely testable, for we can try to test if constants of nature have evolved since the Big Bang. Others have raised similar questions, with tentative evidence for the evolution of the fine structure constant (Barrow et al. 2002). If so, some variations may be late, for there is a very special resonance in three helium atoms that allows carbon to be formed in stars, and tuning of constants and other properties such as masses may have occurred near that time such that carbon and the more complex atoms could be formed in nucleosynthesis, then leading to complex chemistry. This helium resonance is the "put-up job" of Hoyle, mentioned by Paul Davies in *The Goldilocks Enigma* (2006), who also imagines that there can be quantum variation in the laws and constants. The sense of the put-up job is part of what led to the awareness that the constants are very finely tuned to allow our complex universe and physicists to wonder at those constants.

But if the values of the constants, details of the laws, and masses of particles, can evolve, perhaps as described earlier, we are not *forced* to consider the anthropic principle as the only answer. Our single universe may have become as it is by the evolution of the laws, constants, and features of particles such as mass. With that evolution of the laws "selected" for creating via purposeless teleology more opportunities or ways for the cosmos to become, just as in the biosphere, our specific laws are *foundationless*!

We arrive at a conceivable view of the world in which the laws and constants evolve; hence, there is no *fixed foundation*! Don't cry. If the biosphere can become the most complex system we know in the universe without foundations in entailing law, reductionism as a whole must fail, and it is not a tragedy if the universe also becomes without fixed foundations. We just have the fun of new ideas beyond the Pythagorean dream that are also testable.

Can the Laws Emerge from the Higgledy-Piggledy?

Such was a hope of John Archibald Wheeler for quantum mechanics (1990). Well, maybe. Imagine something like "the Possible" outside of space but inside of time that might have predated the Big Bang, as I discuss at the end of this chapter with respect to Hawking's boundryless universe (Hawking 2003), and Penrose's comments in *The Road to Reality* (2004).

If Wheeler is right, why do we have the Schrodinger equation? Why the standard model of particle physics? Some time ago in thinking about this I realized that the Schrodinger linear wave equation generates more diverse solutions, for the sums and differences of solutions are also solutions, than almost any other known equation. This is a topic only slightly explored mathematically, but it is open to investigation. In any case the Schrodinger equation can generate a vast array of solutions given any fixed boundary conditions, and over the enormous array of such boundary conditions such as arose in the evolution of the cosmos we live in. Thus, from the higgledy-piggledy where somehow "laws" are tried, the Schrodinger equation that is deeply interesting also arises in the standard model. Here there is a fixed set of three families of fundamental particles that form groups that transform into one another and only one another. Imagine instead that the particles transformed into ever-new particles in a kind of particle "jet." Such a jet is not inconceivable logically and is a "free algebra" mathematically. But what does a group give us? It is a kind of repeatable fixed set of building blocks, like Lego blocks, from which more complex entities and processes among them can arise and build upon one another to form a complex universe. The particle groups are like a grammar, enabling complex structures such as atoms and chemicals to arise, rather like words, sentences, and paragraphs. Taken together with the Schrodinger equations having very many solutions, the group feature of particles and the Schrodinger equation "enable" the generation of a very complex universe given tuning of the constants. If we imagine Wheeler's higgledy-piggledy of some perhaps unprestatable set of laws, we can imagine the conceivable emergence of the standard model and the Schrodinger equation allowing a complex universe to arise. On this image, the Schrodinger equation and the group structure of the particles arise not only as "laws," but those laws plus tuned constants enable, make possible, the complex universe, perhaps one whose growth is such that its Adjacent Possible, what can become next, tends to grow, as has the Adjacent Possible of the biosphere and our global economy, where the specific becoming of neither, I have argued, is "governed" by no entailing laws at all.

Where or how might Wheeler's higgledy-piggledy arise? One idea is that if res potentia is *outside* of space but inside of time, then the higgledy-piggledy can have occurred in a Possible before the Big Bang. The singularity theorems of general relativity require a beginning of time at the Big Bang so time does not flow from an infinite past to an infinite future. The singularity theorems apply to the Actuals of general relativity (Hawking 2003). General relativity has no Possibles. But, as Hawking argues in "imaginary time" general relativity space-time becomes Euclidian, and it can have no boundary! Just conceivably, time flow in the *Possible*, a rotation of the quantum wave equation in the complex plane, is a physical realization of "imaginary time." Were we able to make this work, the universe of Possibles and Actuals could be boundryless, as Hawking hopes in some extension of general relativity that includes Possibles, of which Penrose's orchestrated objective reduction (OR) is an example (1989, 1994, 2005). I return to this later in this chapter.

In summary, we love the Pythagorean dream and perhaps there is an entailing final theory. Doubt about the latter arises from the fact that no laws appear to entail the becoming of the biosphere, itself a complex part of the universe. In addition, standard difficulties finding the final theory at least allow doubts that it exists.

One has to kind of like the ideas of emergent foundationless laws and tuned constants without the anthropic principle and an untestable proposal of a multiverse whose universes are in no contact. At least foundationless becoming seems conceivable and we are not forced to adopt the Pythagorean dream. But what would approximately "freeze in" the Schrodinger equation and the standard model? The obvious thought is the formation of our universe in something like the Big Bang, perhaps a foundationless boundryless universe of Possibles and Actuals.

The Classical World?

There appears to be no agreed-upon way to get from quantum mechanics to the classical world. The decoherence program (Zurek 1991) is one try. Stapp is another (2011). It is not the main topic of this book, but it would be very nice were we to understand how whatever the classical world may be, it emerges from the quantum world. Whatever our account of the classical world may be, it must allow the data on Earth for a long evolutionary history with the very rich and enormously diverse nonrandom geological record of fossils for at least 700 million years or much more. Evolution and the accumulation of adaptations require a stable-enough or "shared-enough" classical world. So does organismal embodiment. It is not clear to me how to get this stable-enough shared classical world from either Stapp or the decoherence program. The latter leads to a classical mixed state of probabilities in correlation between a system and its "environment" based on von Neumann entropies. Why does this yield or allow the fossil record? Of course, it might.

But there may be a testable alternative to yield the classical world. Might the quantum Zeno effect play a role? Here is the quantum Zeno effect: If a quantum superposition is measured in some basis, one result can be a single wave function, an Actual fulfilling the law of the excluded middle. Then, slowly, that is, quadratically in time, new wave functions flower, first near, then further from the initial measured wave. (So the new Actual of the measured, now Actual, single wave function enables new wave functions or possibilities to flower.) If the system is remeasured rapidly, presumably in the same basis, before the amplitudes of the new waves are large, it will almost certainly be measured to be in the initial wave function, so it is "trapped" in that wave function by rapid remeasurements. Indeed, if remeasured infinitely fast, the propagation of the Schrodinger wave equation is stopped, but infinitely rapid remeasurements cannot happen.

Pauling's theory of the chemical bond requires *single stable wave function per bond*, say between electron and proton exchanging virtual (or real) photons. But this exchange constitutes measurement on the general argument from decoherence (Zurek 1991) that the quantum environment measures the open quantum system. After measurement when the system falls to a single wave function, as noted earlier, it flowers new amplitudes quadratically in time, the flowering can be cut off by rapid remeasurement, the quantum Zeno effect. But then "Why does Pauling's chemical bond remain a *single* wave function, rather than flower new amplitudes?" One answer may be the quantum Zeno effect itself, as say, electron and proton exchange virtual or real photons and remeasure one another the same way rapidly enough to induce the quantum Zeno effect. Then Pauling's bond as a *single wave function remains a single wave function due to the quantum Zeno effect*! This might be testable. And the classical world may arise due to it.

Then the central idea I propose is that quantum variables co-measure one another rapidly and induce the quantum Zeno effect, thereby co-trapping one another in a tiny subset of their possible states and nearly stopping the propagation of the Schrodinger wave. I hope this can lead to a classical enough world, as described in more detail below. But note that if confirmed, co-trapping would be evidence that quantum variables do measure one another.

There is surely at least one major problem: In what basis is the measurement? See Barad (2007) for a superb discussion of Bohr's ontological views denying our capacity to ascribe ontological status to, for example, an electron when we cannot measure both position and momentum with the same apparatus simultaneously. Note that Bohr's position denying ontological Actuality to the electron is yet, I think, entirely consistent with ascribing res potentia, ontologically real *Possibles* status to the unmeasured electron in a dualism in which Reality consists both in Possibles and Actuals linked by measurement, which may be Mind measuring, my Triad. But in what basis is measurement, by mind or apparatus (for Bohr)? Measurement mathematically (von Neumann [1933] 1955) is a projection from the circle in the complex plane that represents the Schrodinger equation wave oscillating, to the real line by operators that make this projection, which projection is called a "basis." There is, in quantum mechanics, no account of which basis is "freely chosen." That for Bohr is the choice of which apparatus. But the choice matters, for what becomes Actual depends upon the basis. For example, if the chosen apparatus measures position, momentum is spread out, and conversely. Perhaps, basing my try at the classical world by the quantum Zeno effect, the choice of basis can be answered by rapid successive measurements in many bases, say at random, thereby holding those all those bases in a narrow range. I note that "Einselection" in the decoherence program hopes to select two bases, position and momentum (Zurek 1991). (The quantum Zeno effect may be involved in decoherence on some views.) We lack an agreed-upon theory of basis choice. Perhaps different quantum variables and forces, let alone classical devices, eyes and ears can only measure in specific different bases. If so, and consciousness is involved, the qualia may or must vary with the basis of measurement.

The idea of co-trapping may be testable. Quantum Zeno effect co-measurement and co-trapping in a single wave function could occur testably in wider contexts. A possibly testable case might be with few versus many electrons and photons in a reflecting cavity creating discrete energy levels, with the real photons chosen to match that needed for transitions between only two nearest adjacent specific mid-energy levels. Such exchange may measure rapidly enough to co-trap the participants in the two neighboring energy levels for a long time. Then might such co-trapping in two mid-energy levels for a period of time longer than the expected on Poisson in time jump statistics be testable? If found, we would have starting evidence that the quantum Zeno effect can co-trap a system for long periods in a few energy levels, or also from Pauling's theory, in a single wave function forming chemical bonds.

Can the quantum Zeno effect explain classicality? Maybe. Much of the classical world is made of molecules with covalent bonds and other bonds or other materials with bond structures. Classicality is held to increase with "mass," so a huge lump of coal is supposed to be more classical than a sugar molecule. We may be able to test for a possible role of Zeno co-trapping increasing with the mass of the material, hence the number and rate-density of possible measurement events in the mass. The idea may well be wrong, of course. But, if testable, and we do not yet know how the classical world arises, it may be worth a try. If the aforementioned were true, the moon would be there when no one is looking because its quantum variables co-measure and hence co-trap one another and prevent the Schrodinger equation from propagating, and so also the moon localized in its orbit when no human is looking, because co-measurements by quantum variables render the moon "classical enough," but still the quantum variables observing and measuring one another would hold the moon in one place in its orbit, not smeared around that orbit. And classical measuring devices and mica could, at the quantum level, carry out quantum measurements and leave "records." As noted earlier, von Neumann is left with a fuzzy range of where to place his epistemic cut, at consciousness or the indefinite sequence of measuring apparatus (von Neumann [1933] 1955), despite the infinite regress of classical devices becoming entangled with the quantum system and one another indefinitely. On this hope, the quantum variables at the base of a quantum Zeno effect classical enough world do the conscious measuring of one another, answering von Neumann, the debated location of measurement, the role of the "classical-enough" apparatus, and also the enigma.

There is in fact, if repeatable, early evidence for a role of the quantum Zeno effect in what is being called "emergent classicality" (Patil et al. 2015). These workers recently used atoms which form a lattice at nanokelvin temperature scales illuminated weakly or strongly with a laser, and they observed a transition to "emergent classicality" as laser intensity was increased and atoms stopped tunneling between locations in their lattice structure. Laser observing the atoms stopped their motion. A role for the quantum Zeno effect in the emergence of a classical enough world may merit further investigation.

Further potential evidence for a quantum Zeno effect role in the emergence of a classical enough world is found in work by Rebentrost et al. (2009), who studied the roles of coherence and decoherence in quantum transport in light harvesting molecules and model systems, finding that at an intermediate level of both, transport is maximized, but at high decoherence, that is, dephasing, the quantum Zeno effect, the "watchdog" effect, arises freezing transport.

A Bridge Too Far?

My main purpose in this book is not physics itself. I am not a physicist, caveat lector once again. Yet I wish now in the name of at least intellectual curiosity, to explore briefly some of the possible implications of what has been said.

Quantum mechanics, including quantum field theory, which has united the electromagnetic force and weak force, but has not yet united these with the strong force, remains un-unified with general relativity. Einstein spanned from special relativity in 1905, with which quantum mechanics is consistent, to his triumph of general relativity in 2015. Schrodinger wrote his equation in 1927. No effort to unite our twin pillars of twentieth-century physics has yet succeeded. Two strands in the effort to unite quantum mechanics and general relativity are "loop quantum gravity," which concerns the very creation of space itself, and string theory, which assumes a background space-time of many dimensions. Neither has yet succeeded. As Leonard Susskind (2006) writes in *The Cosmic Landscape*, there are now an estimated 10^{500} potential and different string theories and no one has written down the equations for any of them. Susskind posits 10^{500} pocket universes, each with its own version of string theory, some more life friendly than others, hence *The Cosmic Landscape*, a version of the weak anthropic principle. Note that by this chapter, we are also not *forced* to the anthropic principle; the constants and laws can have evolved.

The hold of the Pythagorean dream is powerful and in some sense feels deeply right. We should dream of a final and fixed entailing theory, and we should try very hard to find such a theory, with the history of Newton,

Einstein, and quantum mechanics so magisterially before us. On the other hand, failure since 1927 may be a clue that the dream is merely that, a dream. Smolin in *Time Reborn* (2013) remarks that many in his generation of physicists feel they have not accomplished anything of note, unlike the earlier generation that led to quantum field theory and the standard model. It just may be that the failure to unite quantum mechanics and general relativity is a real failure, part of a general failure of reductive materialism. Failure for a long time is at least weak evidence. A wonderful cartoon years ago in *The New Yorker* showed a group of middle-aged men and women lost in a dense forest. A woman has climbed a tree and is peering out. "Wrong forest!" she calls down to her companions.

I want to suggest we look at the possibility of "wrong forest."

Might it be the case, if Possibles are ontologically real, but the classical physics in general relativity has no Possibles in the theory, that we have difficulty uniting quantum mechanics, and quantum field theory with general relativity because general relativity lacks Possibles? Then if we can somehow get "the classical" world of Actuals from quantum mechanics, for example, my try earlier with the quantum Zeno effect, perhaps we do not *need* to unite quantum mechanics and general relativity by quantizing general relativity at all!

But even were I right about a classical world via the quantum Zeno effect, that world is in the pregiven background classical space-time of general relativity; thus, a quantum Zeno effect account of a classical-enough world offers no account of classical space-time itself, let alone the mystery of the Big Bang itself, and even its origins. Tries to quantize general relativity via the hopes of loop quantum gravity and string theory still have not succeeded. Perhaps we do need to quantize general relativity, but the dream to quantize general relativity in these ways is "the wrong forest."

Then perhaps there is a different forest that may yet quantize general relativity, hence space-time itself, in a new way. Hawking wants a boundryless universe, but that requires imaginary time, which converts the metric of general relativity from (−+++) to (++++), that is, to Euclidian. Such a Euclidian space-time can form a closed four-dimensional sphere, hence have no boundaries, hence Hawking's dream. Remarkably, Penrose (2004), in discussing Hawking's idea, notes the following: (1) Imaginary time is needed for consistent quantum field theories. (2) Imaginary time is needed for positive frequencies. (3) Unlike path integrals in a *prestated space* or *space-time*, Hawking's proposal of *imaginary time* requires that *each path be its own separate space-time*. In turn, claims Penrose, this requires that *these different space-times are in superposition.* This is why Hawking suggests that this is the right way to quantize general relativity. But if, by my earlier arguments, superpositions do not obey Aristotle's law of the excluded middle, and by C. S. Peirce *Possibles* do not obey the law of the excluded middle, then we can consider the idea that the superpositions of different space-times, one per pathway, are different *possible space-times*! But this step does seem to quantize general relativity in a new way, new forest! Furthermore, here there *is* no prestated single Actual background space-time, as in string theory. Also, given a set of *possible* space-times, Possibles are indeed *outside of a single classical space-time*, as I suggested might be the case in Chapter 7. Penrose then suggests that at the Big Bang, most or all of the superpositions of space-times would vanish somehow (via measurement?), converting back to a GR metric (−+++), but that some Euclidian space-times flowing in imaginary time might persist indefinitely, allowing time to flow, but as "imaginary time," with metric (++++). This could give Hawking his finite but boundryless universe, consistent with Possibles outside of actual space but inside of a boundryless imaginary time in the boundryless universe. We have no interpretation physically of "imaginary time." What if time flowing in the Possible is imaginary time (Katherine P. Kauffman, personal communication)? Perhaps time flows *only* in the Possibles as the fundamental, imaginary time, just as Hawking wonders, and if Possibles are outside of a single classical space-time but inside of imaginary fundamental time, we can think of what happened before the Big Bang without a background space-time. If the Actual universe and Big Bang emerged from the Possible in this boundryless universe, then perhaps the Actual Big Bang universe comes to exist from the *possible, thus from nothing actual*, a new way to ask: "Why something rather than nothing?" The answer from the higgledy-piggledy becomes: "Because the universe *could* become, and perhaps because it could become complex." Perhaps we can get foundationless laws from the higgledy-piggledy, as Wheeler hoped, arising in the Possible before the Actual Big Bang, and a finite boundryless universe á la Hawking, of both ontologically real Possible and Actuals, even perhaps linked by Mind measuring in the Possibles. Then if general relativity is purely Actuals, and no time flows among Actuals, time can become merely a *dimension* that does not flow in any single actual space-time of general relativity that is the pure causal structure among Actuals in a fixed background space-time. Penrose speculates that imaginary time and superpositions of space-times can persist, thus perhaps throughout Hawking's finite universe. Perhaps they do, imaginary time flowing, and perhaps they can serve as a basis of quantum variations in the laws and constants, as Davies imagines (2006), and I do in this chapter.

More measurement of entangled variables can yield "*now*" for that set of variables whose single wave function changes instantaneously upon measurement of any of the variables. Penrose's OR (Penrose 1989, 1994) is one means to choose on a Planck scale between different possible space-times. Even on OR, the universe must be dense with Penrose space-time possibles bubbling and collapsing, and perhaps entangled. Perhaps mind measures and collapses, rather than OR with a burst of consciousness and free will, which seem on OR, a rather ad hoc addition of the mental into a purely physical universe.

Is this the Pythagorean dream? Perhaps, but without fixed laws and constant required, and it still does not entail the sufficiently classical world allowing the evolution of the biosphere or all of life and the vast diversity of the fossil record.

Conclusion

This chapter has explored the idea that we need not be Pythagoreans, nor believe in a fixed foundational theory of everything that entails all, which, if the biosphere becomes beyond entailing law and is part of the universe, must fail in any case. The ideas are highly speculative, perhaps not *even wrong*, but some are testable and were they found true would transform physics. Perhaps we are without fixed foundations. Perhaps this may become a useful idea. We are trapped in the Pythagorean dream, but the higgledy-piggledy may be possible. The purpose of this chapter is achieved if you are persuaded that we are not forced to assume fixed foundations. If there are none, our universe of Possible and Actuals is emergent and enabled by its current but changing Actuals, and we live in it.

Who Are We Knowing, Doing, Living Humans in a Creative Universe in Late Modernity?

Part III now hopes to begin, begin only, to explore what we are as humans in a creative universe, we in late modernity or postmodernity. A stage was set in the Axial Age, more than a thousand years after writing was invented, from 800 to 200 B.C.E. with the advent of our classics across the globe, Confucius, Buddha, Amos, Jeremiah, Isaiah, Plato, and Aristotle. All sought in diverse ways, forms of transcendence beyond early religions that sought welfare for the local community, good hunting, and nourishing rains, by propitiating the early gods. Transcendence was and remains a search for something higher—ethics for Confucius, enlightenment and compassion for all sentient beings for Buddha, and the good, the true, and the beautiful for Plato. Yet the transcendent is also somehow beyond statement. In the allegory of the cave, Plato describes us in a cave seeing only shadows. Turning to the sunlight we glimpse *"the good,"* never defined. Were we to look too long, we would become blind. The allegory of the cave is a metaphor that evokes and inspires without precisely stating. We have substantially lost transcendence and enchantment in late modernity. Yet we really are perhaps at the most enormous one of Thomas Cahill's "hinges of history," in which our thirty ancient civilizations around the globe are weaving together, affording us enormous opportunities to co-create what we will become. If Parts I and II rend the overtaut weave of fine science that binds us in the hegemony of the Reductionist dream, thus may set us free, Part III would hope that, free, we can dream anew.

Summary of Parts I and II

The first two parts of this book have sought to set us free from the hegemony of reductive materialism in which all that becomes in the universe is entailed, whether in classical physics or quantum physics. Chapter 3 points to an antientropic process above the level of atoms where the universe is nonergodic and in part underlies all that is history becoming (but respects the second law); Chapter 3 also claims reductionism fails even in classical physics and chemistry. Chapter 4 on the evolution of the biosphere claims no laws entail the specific evolution of the biosphere and introduces the new ideas of the Adjacent Possible and Actuals, which are enabling constraints that do not cause, but enable, new Adjacent Possibles, typically unprestatable, into which the biosphere evolves. More natural selection does not "act" to achieve the very possibilities for future evolution that arise. The biosphere flows, drawn into some of the possibilities it creates! This is an entirely new way of thinking, involving enablement, not cause, but enabled created by Actuals in the world, the swim bladder for example, and being "sucked" into some of the very possibilities evolution itself creates. This is a new idea which I think applies, as shown in Chapter 5 on innovation in the economy, to economic development since 50,000 years ago when we had only perhaps 10,000 goods and production functions on the globe and now have 10 billion in New York alone. We have exploded, as has the biosphere, into the Adjacent Possibles we largely unknowingly create. Here is Nagle's "purposeless teleology" (2012). The greater the diversity of things and processes, the greater the ways they can become Actuals that enable new Adjacent Possibles into which life becomes in prestatable and unprestatable ways. Chapter 13 (a sort of addendum to Part I) shows that, if quantum measurement is real and ontologically indeterminate, then the evolution of the world at the quantum level is a *history* of what happened that is not entailed, and sometimes not even the Adjacent Possible is prestatable at the level of ever-new quantum behaviors enabled by sometimes unpredictable actual boundary conditions from prior measurements of chaotic quantum systems. Where true, at this quantum level, the history is unpredictable ab initio, new in the universe Adjacent Possibles arise, and parallel the evolution of the biosphere and the economy and other aspects of human life. Chapter 14 explores whether we are forced to hold the Pythagorean dream of foundational laws outside the universe. I conclude that we are not so forced. Reductionism for the entire universe will fail to entail all if the becoming of the biosphere is unentailed and part of the universe. But Chapter 14 envisions the evolution of laws and constants, which act, via that evolution, as enabling constraints as well as causes. Part I claims we are in a *creative universe.*

Part II, after the failure of Descartes' res cogitans and the triumph of his res extensa in Newton and classical physics with its causal closure and at best an epiphenomenal mind and no free will, seeks our subjective pole, with all our sense of a conscious mind that can sometimes act with responsible free will in the world. I hope my arguments are cogent, sometimes testable, but not entirely testable. I break the causal closure of classical physics with some confidence using quantum mechanics and acausal consequences of a quantum coherent and Poised Realm mind for the classical meat of the brain and body via acausal decoherence and recoherence, and presumptively acausal quantum measurement. I propose the testable hypothesis that measurement is necessary for consciousness. I find hopeful grounds for free will for humans and electrons in the strong free will theorem. And in suggesting a solution to the quantum enigma, I find grounds for a free will in the reality of ontologically indeterminate measurement so the present could have been different. I propose that quantum variables consciously measure one another, and consciously and responsibly decide with free will. Thus, I extend the enigma from us to particle, a step that has the virtue of saying that measurement can occur anywhere on von Neumann's range from consciousness to the apparatus measuring via its quantum variables measuring one another. I then try in the Triad to unite Actuals, Possibles, and Mind, where Actuals obey the law of the excluded middle; Possibles, quantum superpositions for example, do not; and Mind measures Possibles to yield New Actuals that yield new Possibles for Mind again to measure in a panpsychist participatory universe. There is even some hope that by the quantum Zeno effect stopping the propagation of the Schrodinger waves, such cross observation can yield a classical-enough world. If Part II is correct, we can have our minds back.

We can have, with evolution building upon this from the minds of electrons to early life to us, our full humanity back, consistent with science.

15

Self-Organization in the Origin of Life, Agency, and Ontogeny

OUR HUMANITY DERIVES from the origin of life on Earth or elsewhere. In the present chapter I begin by telling you what we now know about the origin of life, an unsolved mystery, the onset of agency, and some aspects of knowing how to reliably act in the worlds life creates. Given Part II, it is open to us to consider that even simple life has rudimentary consciousness and free will. Perhaps yes, perhaps not. I will keep both options open, for if mind is part of the universe, it did not originate with life, but it may well have been part of life from the start and, if so, must have evolved enormously to yield the human mind, cognitive, emotional, intuitive, sensate, including a presumptive evolution of some of our morals in the Paleolithic, even if such an evolutionary theory does not capture all that we have become with language and culture.

Andreas Weber (2013) emphasizes that we must re-vision ourselves as primarily *alive*! Well, we are *alive*. On the ceiling of the Sistine Chapel, Michelangelo created one of the most powerful images in the Western world: God's finger outstretched and almost touching Adam's finger, his arm extended to almost touch the hand of God. We sense the spark that will pass from God to Adam. Is the spark the old concept of vitalism? Does life have some mysterious force not present in the living world? Is the spark spirit, mind, or soul? Given the ideas in Part II, we can begin to imagine the spark is spirit, mind, or soul. Do we need the vitalism with which life was viewed in past centuries? No and perhaps a qualified yes. You see, we still consider the origin of life issue one of classical physics and chemistry, and ask, "Whence alive?" It may be as I discuss next that classical physics and chemistry suffice for the origin of life—I too have contributed here

(Kauffman, 1971, 1986, 1993). But Part II raises the possibility that far more than classical physics may be hinted at in quantum mechanics, with quantum biology now exploding (Engel et al. 2007) and the possibility that mind and free will are part of the universe itself, hence of life. Is this God's finger almost touching Adam's? In truth, I suspect it could be, for a knowing, doing universe is so far beyond our seventeenth-century ideas of matter, and nineteenth-century ideas of energy, that, Horatio, we must wonder. What is it to be alive, and living, to have evolved to advanced life forms, including our human aliveness? Surely our aliveness is not captured by science itself, even as I sketch the science behind the current attempt to understand the origin of life. Life is living. What will we do with our aliveness with one another and nature? Shakespeare and Pushkin are immanent in life to touch God's finger.

The Origin of Life Problem

Before Pasteur, there was no problem of the origin of life. New life sprang from each rotten log after a rain when maggots appeared. But Pasteur won a prize by using a beaker containing sterile water; the beaker had an S-shaped neck with water in the bottom of the S, which blocked atmospheric air from entry into the flask. This demonstrated that no bacteria grew in the sterile flask. Life, he said, comes only from life.

I stress that all the major ideas discussed herein are based on classical physics and chemistry. However, the self-organization I will discuss with respect to the spontaneous emergence of "collectively autocatalytic sets" (Kauffman 1971, 1986, 1993) is, in fact, independent of any specific classical physics and chemistry, for the same ideas apply to the evolution of the economy. I will mention the possible roles of the Poised Realm and quantum behavior later. Without them, life is a classical physics machine, incapable of choice where the present could have been different and incapable of more than an epiphenomenal mind. We are, however, given quantum biology now emerging, no longer forced to limit ourselves to classical physics and chemistry.

With that, the problem of the *origin* of life arose. Darwin does not presume to solve the origin of life; he starts with life and discovers evolution by natural selection, the most stunning modification in our Western thought since Newton.

No progress was made on the origin of life until the early twentieth century, by Oparin and Haldane. Oparin (1968) proposed that life started as coacervates in liquid medium. Haldane proposed the "primordial soup" (1954), a dilute mixture of the atoms of organic chemistry, carbon, hydrogen, nitrogen, oxygen, phosphorus, and sulfur (CHNOPS) and a variety of small organic molecules. Life would somehow arise from this soup, often shown in origin-of-life talks as a Campbell soup can labeled "primordial soup."

The next major step was taken by then graduate student Stanley Miller in 1953. He made a beaker containing water and simple molecules such as methane, ammonia, and others, which were thought to be the reducing atmosphere of the early Earth. He then zapped these with electric sparks to mimic lightning and to add energy to the system. After some days, Miller found a brown sludge on the bottom, and it contained about five of the biological amino acids. Abiotic synthesis of parts of the primordial soup had been shown experimentally. This led to decades of work on prebiotic chemistry with synthesis of sugars, nucleosides, amino acids, and lipids. The field flourished, but as was noted, the conditions to generate one compound were different from the conditions for another compound, leaving open how these might assemble in a single small region.

Then we discovered that chondronacious meteorites, that is, containing CHNOPS, such as the Murchison meteorite that fell in Australia in the 1970s, contained thousands, indeed at least 14,000 different organic molecules at the femtomole range. This and other meteorites were formed as the solar system was forming, and there was a long and vast infall of these chondronacious meteorites, we think, onto the early Earth and its early oceans, creating more primitive soup.

Life requires energy. Sources of energy include solar photons, lightning, hot thermal vents in the ocean floor, sulfur compounds, and others. Recently bacteria living only on electrons and their flow have been found.

All this is hopeful. But where did molecular reproduction come from? Here there are a number of competing ideas and experimental traditions. Note that in some sense molecular reproduction and protocells require a classical-enough world and Anderson's generalized rigidity to be embodied.

The RNA Worldview

In 1953, Watson and Crick discovered the double-helix structure of DNA and, with British understatement, say at the end of their paper, "It has not escaped our attention that the structure of the molecule suggests how it might be replicated." Indeed, it was soon shown that one side of the helix, call it Watson, separated from the other side of the helix, call it Crick, and due to the specificity of base paring A with T and C with G, the sequence of A, T, C, and G on the "old" Watson strand does specify the sequence of A, T, C, and G on the newly synthesized Crick strand. Similarly, the old Crick strand specifies the sequence of A, T, C, and G on the newly synthesized Watson strand. So, with the help of a number of enzymes, DNA replicates rather as Watson and Crick hinted.

As many of you know, in contemporary cells, DNA sequences are transcribed into complementary RNA sequences, where RNA is a polynucleotide cousin of DNA. Then via the genetic code and protein enzymes and the ribozyme, the RNA can be a messenger RNA that is translated into a linear linked sequence of amino acids linked by peptide bonds, to form the primary sequence of a protein that later folds and performs some cellular function. I stress again the magic that translation requires for proteins that carry out translation and are self-consistently encoded by the code that specifies their very sequences. The origin of this code and translational system, which is self-consistent, remains a huge mystery.

RNA can form double helices, as can DNA. It was obvious that RNA might be a basis for the origin of life. The basic idea was to take a single strand of RNA, say the "Crick" template strand, in some buffer, but without evolved protein enzymes; add the free nucleotides of RNA, A, U, C and G; and hope that Watson-Crick base pairing would line up the free nucleotides on the single-strand RNA "template." Thus, if that template were "AUCCGUA," it would line up free U, A, G, G, C, A, U nucleotides and then link them together with proper 3′-5′ phosphodiester bonds to create a complementary strand, the new "Watson strand." Then the two strands were to melt and separate and the process was to repeat, creating self-replicating RNA Watson and Crick strands. Note that the chemistry of DNA and its replication depend upon the specific physics and chemistry involved, which at the classical level are efficient causes.

Brilliant organic chemists, led by Leslie Orgel at the Salk Institute (Orgel and Lorhmann 1974) and many others, have tried for some forty years to get this to work with RNA, and cousins such a PNA. All efforts have failed so far. The reasons are partly known. RNA building blocks, nucleotides, have numbered carbons on their ring structure. Pairs of RNA nucleotides thermodynamically favor forming 2′-5′ bonds, which do not fit into strands making a double helix, which requires 3′-5′ bonds. Sequences of CCCCCCC can be made to yield sequences of GGGGGGG, but the GGGGGGG sequence folds and precipitates. The approach may yet work without enzymes.

Then a stunning discovery was made. It was known that proteins often function as enzymes, speeding chemical reactions in their approach to chemical equilibrium. Then it was discovered that RNA sequences, now called ribozymes, could also act as enzymes. Biologists were thrilled. Somehow RNA molecules could carry genetic information and also act as enzymes that might help replicate single-stranded RNA sequences. The RNA *world* was born a few decades ago. I confess I have always been puzzled by the idea that RNA carries "genetic information" when, as noted earlier, translation into proteins of messenger RNA requires self-consistently encoded proteins to mediate the translation by matching the right amino acid with the right RNA triplet codon. Without encoding, an RNA sequence is just an RNA polymer and carries no genetic information.

Efforts to evolve such a ribozyme from libraries of random RNA sequences, which helped replicate single-stranded RNA, acting as a polymerase that adds nucleotides one by one to the second strand being synthesized by the first strand template, have failed to get beyond adding some fourteen nucleotides. This work seems to have halted but may yet succeed. How such an RNA polymerase might form prebiotically is unclear, and whether it would be evolutionarily stable as it reproduced itself with some errors, and the error-laden copies made yet more copies with even more errors, is not clear. A runaway "error catastrophe" of the Orgel type where the error rate increases as polymerases with ever more mutants are made, might occur, and this issue has not, to my knowledge, been investigated.

How to Get a First RNA Single Strand Has Been a Problem, Perhaps Recently Solved

In the RNA world, which is dominant in NASA, further evidence is that ancient co-enzymes are RNA, and some begin to think of RNA and small proteins, peptides, working together early in the origin of life. Ada Yonath, Nobel Prize winner for work on the ribozyme, which is the center of protein synthesis, wants to hope that that RNA molecule can be reproduced and stabilized by a subset of the peptides it makes to create a functional whole (Yonath 2014).

That is roughly the evolving current status of the RNA world. It too depends upon specific efficient cause chemistry in classical physics and chemistry.

The Lipid Worldview

Your cell membrane is made of two layers of lipids, each with a polar and nonpolar head, the first water loving and the second water hating. Such "bilipid layers" form spontaneously and easily for many lipids if each at sufficient concentration is merely dissolved in water, perhaps with some buffer. A bilipid layer is formed that then closes into a hollow vesicle called a "liposome." Wonderful experiments by Luigi Luisi (Stano, Wherli, and Luisi 2006) at the E. T. H. in Zurich and David Daemer at the University of California at Santa Cruz (Daemer and Barchfeld 1982) have shown that liposomes can grow by adding lipid and *bud and divide*. These then form molecular self-reproducing systems. Furthermore, they can absorb DNA, RNA, and other molecules if driven in the presence of these DNA, RNA, and other molecules, through wet-dry cycles and by other means. These results are very hopeful; we have a form of molecular reproduction, but only of liposomes or their cousins, lipid micelles. Again, the lipid world depends upon specific efficient causes in classical physics and chemistry.

The Uncatalyzed Chemical Autocatalytic Cycle Metabolism Worldview

Your cells and bacterial cells have a catalyzed chemical cycle, the TCA cycle at the base of metabolism. In the absence of enzymes this cycle can run backward, and in doing so in the presence of its building blocks, it produces two copies of one of its members, thus creating a chemical autocatalytic loop that makes a second copy of one of its members. It remains a still largely unexplored issue in complex chemical reaction systems how many autocatalytic chemical reaction loops form. Do they interweave in interesting ways, perhaps producing many second copies of many compounds?

Uncatalyzed Chemical Reaction Cycles Might Synthesize Molecules Able to Act as Catalysts

Recently Richard Behrens (personal communication 2014), who studies explosive materials, told me that a largish explosive molecule broke down into a smaller molecule, NO, which then formed a chemical autocatalytic loop with the larger explosive, breaking it down again to produce more NO. But, noted Behrens, the larger molecule has peptide bonds and some of the explosion reaction products are linear and branched peptides. This has just led Behrens and myself to wonder if chemical autocatalytic loops might also synthesize larger molecules, say peptides as earlier or others, that could act as true *catalysts* and speed some or all of the chemical reactions, which thereby form a catalyzed metabolism. In turn, might these fulfill an old idea of mine, discussed later, of the emergence of collectively autocatalytic sets of polymers, such as the nine-peptide collectively autocatalytic set of Gonen Ashkenasy at Ben Gurion (Wagner and Ashkenasy 2009), mentioned in an earlier chapter and later in this chapter.

In 1971, three independent papers appeared that sought to find a solution to the origin of life.

The first paper, by Manfred Eigen (Eigen 1971; Eigen and Schuster 1978), in Göttingen at the Max Planck Institute, proposed sets of pairs of + and − nucleotide strands as in the Orgel experiment, but each such pair, which would replicate as an autocatalytic cycle, would catalyze or help the next such pair to replicate around a hypercycle of these RNA replicating strand pairs. An enormous body of theory now surrounds the hypercycle model, which has its defects. For example, if the hypercycle has more than three members, it tends to fall apart into shorter hypercycles with three or fewer members. Since no one has succeeded in Orgel's experiment, there are no experimental tests of the hypercycle model based on template replication of RNA, DNA, or other polynucleotide sequences as yet. Because the hypercycle depends upon sets of template-replicating RNA sequences, it too depends upon efficient causes in classical physics and chemistry.

The second paper was by Timor Ganti (2003) of Eotvos University Budapest. Ganti proposed the first full-scale model of a minimal cell, the Chemotron, containing a minimal metabolism to supply energy and building blocks, a replicating RNA molecule, and surrounded by a lipid membrane like a liposome. It is hard not to like Ganti's model, and Szostak at Harvard Medical School and others are now working with different lipid liposomes containing evolved peptides and nucleotide sequences to see if reproduction can be achieved. But these efforts use evolved molecules.

The third paper was by myself (Kauffman 1971, 1986, 1993). I proposed the spontaneous emergence of collectively autocatalytic sets of polymers, specifically peptides. However, my theory is, in fact, independent of any specific chemistry and even the material and energetic basis of such systems. It is an example of formal cause "self-organization." My own thinking was a bit wacky. I wondered, if the constants of nature were different and we could form chemicals but not A, C, T, G, and U, to form the double helices of DNA and RNA, would life be impossible? Nuts, I thought. Life must be based on the catalyzed, hence, kinetically controlled and reasonably rapid, formation of a set of molecules that catalyzes its own formation from chemical building blocks supplied from the outside. Such a set would be "collectively autocatalytic" in the precise sense that no molecule catalyzed its own formation, but the formation of another member of the set, and such that the set *as a whole* jointly catalyzed its own formation from outside chemical building blocks. In that first paper, I modeled molecules as formally abstract binary strings and assigned which bit string molecule catalyzed which reaction among these strings, gluing them into longer strings, a ligation reaction, or breaking them apart into smaller strings, a cleavage reaction; then a random logical or "Boolean" function was assigned to each reaction with more than one "catalytic input molecule," so some combinations could inhibit the reaction and others could enhance it. My simulations found, as I expected, that as the diversity of molecular model species, my binary strings, increased, and the diversity of reaction among them increased *even more rapidly*, I could find some probability, P, that any model molecule catalyzed any given reaction, at which the formation of collectively autocatalytic sets emerged spontaneously as a "phase transition" (Kauffman 1971, 1986, 1993).

Thus, the emergence of life would be spontaneous and an example of "self-organization" and "order for free," fully independent of the details of what the abstract "molecules" were. Indeed, this is a theory of objects, here molecules, transformations among objects, here chemical reactions, and objects, here molecules, helping or hindering these transformations, here by catalysis. But the theory is independent of the stuff and materials and energy, and I will later show you that an economy is also a collectively autocatalytic set among input goods transforming to output goods, mediated by production functions which use products of the economy to mediate the production functions. Thus, this is a new kind of theory, independent of any specific efficient causes. Indeed, if free will operates in the economy, we may not want, given Part II, to think in terms of efficient cause at all. I think this example of self-organization, independent of the stuff and energy sources, is a "formal cause" law. It is an example of a law of spontaneous self-organization independent of the material basis. If this is right, the origin of life does not require a new law of "physics," but rather a new kind of law, a formal cause law.

This work became a set of theorems, first by myself in 1986, and in simulations with D. Farmer and N. Packard (1986) where we generalized from peptides to RNA sequences as well. Recently fine theorems by Hordijk, Steel, and colleagues (Hordijk et al., 2010, 2011, 2012; Mossel and Steel 2005) have vastly improved my early work. These authors call my collectively autocatalytic sets (CASs) RAFs (an RAF is an autocatalytic set in which reactions from the food set are also catalyzed by the members of the autocatalytic set) for they demanded that the reactions from the exogenous "food set" to the autocatalytic set themselves be catalyzed. I thought a CAS was one vast autocatalytic set. I was wrong. A RAF can be broken into irreducible RAFs, where removal of most single molecules kills autocatalysis, but a set of catalyzed reactions can "hang" off the core of the irreducible RAF. These irreducible RAFs can be combined into more complex RAFs in many alternative ways, forming a partially ordered set among these combinations of irreducible RAFs. Furthermore, criticism of my model had shown that my P assumed that as the numbers of molecules increased, the number of reactions each catalyst catalyzed increased exponentially, which is silly. Hordijk and Steel fixed this. The number of reactions a catalyst needs to catalyze ranges between 1 and 2, which is fully reasonable.

Experimental Work

The CAS theory rests on the distribution of catalytic properties among a family of molecules, say random peptides, RNA sequences, or both, acting on one another. Cloning was invented in the late 1970s, and at some point I realized one could test this question by making libraries of random DNA, RNA, or peptide sequences and testing these for catalytic or other properties such as reliable folding into three-dimensional structures or acting as ligands to bind other molecules. Indeed, it seemed obvious that one could seek a peptide that bound the estrogen receptor and hence mimicked or antagonized the action of estrogen. A drug could be sought. My colleague Marc Ballivet and I at the University of Geneva made the first such libraries and obtained a patent on these ideas filed in 1985 and widely issued (Ballivet and Kauffman 1987, 1989, 1990; Kauffman and Ballivet 1998). This was one foundation for what became "combinatorial chemistry" in the drug industry. Thomas LaBean in my lab (LaBean et al. 2011), and later, Luigi Luisi (Chiarabelli 2006), showed that random peptides could fold reliably. Indeed, Luisi finds that about 30 percent of such peptides fold. If folding is required for function biologically, it is not rare.

In 1995, Reza Ghadiri at the Scripps Institute published a seminal paper. He made a short 32-amino-acid-long peptide that could ligate a 17- and a 15-amino-acid-long peptide fragment of itself into a second copy of the initial

32-amino-acid-long peptide. Peptides can replicate! This established unequivocally that polynucleotides and template replication á la Orgel is *not necessary for* molecular reproduction of chemical polymers. Peptides do it (Lee et al. 1997).

Shortly thereafter, Gonen Ashkenasy in Ghadiri's lab created a nine-peptide collectively autocatalytic set, the one I've discussed before (Wagner and Ashkenasy 2009). This is a Kantian whole; the whole set existed by and for the peptides within it. The function of a peptide is now a subset of its causal consequences: catalyzing the formation of the next peptide, not jiggling water in the petri plate. Functions are real in the nonergodic universe above the level of atoms.

Calling "catalyzing a reaction" a function, Gonen's nine-peptide set achieves *functional* and catalytic closure as a Kantian whole. I used this in Chapter 4.

Ashkenasy has now created peptides in autocatalytic sets that can realize all sixteen Boolean functions on two inputs. The dynamics of such systems can now be studied and can relate to dynamical criticality, and multiple dynamical attractors, discussed later.

In 2012, Nilesh Vaidya and Niles Lehman achieved the next major step (Vaidya et al. 2012). They took a set of evolved ribozymes, cut each in half, separating the catalytic from the recognition site of each, and put the set of fragments in water with only magnesium ions. These two half ribozymes can form hybrid ribozymes spontaneously, with the catalytic site of one now part of a hybrid with the recognition site of the other. They discovered that their system *spontaneously formed single autcatalytic ribozymes, succeeded by sets of three, then five, and then seven* collectively autocatalytic sets of ribozymes! The longer, three, five, seven sets kinetically out-competed the single autocatalytic set.

The spontaneous formation of single and collectively autocatalytic sets of polymers is established.

However, these authors used evolved ribozymes, and Lehman is now pursuing randomizing these sequences away from evolved catalytic and recognition sites. Soon we will know more about the spontaneous formation of collectively autocatalytic sets from random RNA, peptide, or other libraries, or by using Behrens's idea the formation of linked noncatalyzed autocatalytic chemical cycles that then form molecules able to catalyze some of these reaction loops and eventually one another's formation, thus a metabolism with a reproducing RAF inside.

Theory

Eors Szathmary of Eotvos University Budapest, and a student of Ganti, led a group of us in showing that RAFs can evolve to a limited extent (Fernando et al. 2012; Vasas et al. 2012), upon which I enlarge later. In effect, the irreducible RAFs are like self-reproducing "genes," the little tails of reactions hanging off the cores of these are like phenotypes, and an RAF containing many irreducible RAFs inside can undergo selection to lose old polymers and find their way to new polymers. For example, an RAF, if exposed transiently to an exogenous molecule, can "grab" it and incorporate it into a new irreducible or more complex RAF in which the new molecule is now itself produced! RAFs are thus models of simple ecosystems as well that can evolve new "species" of irreducible RAFs.

Roberto Serra (Villani et al. 2014) in Modena has shown that a reproducing RAF in a dividing container like a liposome will have the property that the division of the liposome *synchronizes* with the reproduction of the RAF it contains.

The earlier results start to promise that a reproducing peptide or peptide and RNA RAF in a budding liposome, where the RAF also makes the lipids that are the components of the liposome, allowing it to grow in area and hence divide, constitutes the promise of a *protocell*!

In summary, current work on the origin of life may not be far from the spontaneous formation of protocells, with the spontaneous formation either of RAFs or richly linked autocatalytic chemical reaction cycles. These cycles slowly make molecules able to catalyze both these and other reactions and the collectively autocatalytic formation of the catalysts themselves plus the formation of the lipid building blocks of the liposome membrane, which encloses and confines these so that reactions and reproduction continue to occur, with at least some evolution.

But that evolution can be far wider than merely new polymers. Suppose such a protocell evolves a peptide that buries in the liposome membrane and binds the protocell to a tiny rock in a flowing stream of nutrients. Compared to protocells floating in that stream, the now bound-to-a-rock protocell is exposed to a higher flux of nutrients. It has evolved into a "sessile feeder." A new function, sessile feeding, has evolved as an unprestatable preadaptation! The capacity of such systems to evolve in unprestatable ways is huge, as emphasized in Chapter 4. Open-ended evolution may be easier to achieve than we have thought.

Finally, as in Chapter 4, the fundamental notion is functional closure or sufficient closure of the organism, the Kantian whole in its world with other organisms with which it may interact functionally in very rich ways. What evolve are sufficient functional closures of organisms in their worlds, which collectively constitute the expanding biosphere, which has evolved from a few to myriad species over 3.7 billion years. Where is the "boundary" of the biosphere? We cannot state it; it becomes into its ever-exploding Adjacent Possible that it itself creates. There is no truth beyond magic, it is sometimes said. This is unentailed magic. Furthermore, the selfish gene view is clearly inadequate. Genes are indeed the major means of heritability in current life, but they are useless by themselves. It requires an entire cell, a Kantian whole, to reproduce. And selection acts on these functionally sufficient wholes living with one another in myriad ways as the biosphere evolves. In a sense, DNA, with encoded protein synthesis and enabling constraint that enhances evolution, also goes along for the ride in a co-participatory way.

Agency, Work Cycles, and Maximum Power in Still Classical Physics and Chemistry

The earlier discussion leaves energy entirely out of the picture. In physics, balls roll down hills, but work must be done on them to push them up the hill. The chemical equivalent is that some chemical reactions give off free energy, are "exergonic," and occur spontaneously. Other reactions require the input of free energy and do not occur spontaneously; these are "endergonic" reactions. Real cells link exergonic and endergonic reactions.

An RAF, so far, could be purely exergonic. Now imagine a world in which only exergonic reactions are considered and can interact. Now consider a world with both exergonic and endergonic reactions that can be linked, including exergonic reactions driving endergonic ones. It is intuitively obvious that the latter, mixing linked exergonic and endergonic reactions, is far richer, thus far more able to form autocatalytic chemical reaction cycles and, by catalysis, form RAFs in which the formation of the catalysts is endergonic, as peptide and phosphodiester bond formation is. The synthesis of lipids is endergonic and must be driven to then support the protocell's growth and division.

But it now will become clear that if exergonic and endergonic reactions are linked, the total system must do at least one work cycle; otherwise the system cannot repeat its behavior. In my third book, *Investigations* (Kauffman 2000), I was considering work cycles such as the Carnot engine. I drew an amusing picture of a cannon with powder and ball inside it, aimed at a paddle wheel, which straddled a well. A cord hangs from the paddle wheel and is tied to a bucket down the well and filled with water. The cannon is fired and the ball blasted outward in an exergonic reaction; it hits the paddle wheel, which rotates in an exergonic reaction, which winds the cord up the wheel, lifting the water-filled bucket in an endergonic reaction. The bucket spills over the side into a trough, flows down a pipe, and opens a flap valve to water a bean field. At the end of this, the bean field is watered. The bucket lies on the ground beside the paddle wheel at the top of the well. The cannon ball has rolled into the grass beyond the paddle wheel.

If I put powder again into the cannon, that is, if I supply *energy* to the cannon or *feed* the cannon, can I water my bean field again? *No!* I must fetch the ball and replace it in the cannon, doing work to do so, so an exergonic process. I must lift the bucket, exergonic, and let it fall into the well, endergonic. *Then* I can water my bean field again by firing the cannon again. I have completed the *work cycle*, carried out in real engines by gears, cams, and so forth.

In important summary, if we wish to link exergonic and endergonic processes and repeat whatever they do, the system must complete a work cycle. But then if it is easier for the origin of life and evolution to form RAFs and protocells by linking exergonic and endergonic reactions, these systems must complete work cycles.

Real cells do complete work cycles.

In turn, this raises a central issue concerning irreversible thermodynamics. Here Onsager began the fundamental work about chemical reaction systems displaced from equilibrium, followed by Turing and his model for morphogenesis and Prigogine with his dissipative structures (Nicholis and Prigogine 1977), deeply similar to Turing's ideas.

But in this work there is no concept of an "optimal" displacement from equilibrium. I now suggest an optimal displacement:

Carnot showed that maximum *energy efficiency* of a work cycle such as his heat engine occurs if the work cycle is done "infinitely slowly," or adiabatically. But if cell division of bacteria were infinitely slow, they would *lose* the Darwinian natural selection race! Minimizing energy must be the wrong concept.

Let's try maximizing power efficiency. Consider a car driving and ask at what speed does one maximize miles per gallon? Now "power" is work per unit time. The car driving at constant velocity is overcoming wind resistance and road friction, and doing work to keep moving at that velocity, so it is exerting power. Miles per gallon measures power efficiency per unit fuel. At what speed, 1 mile per hour, 47 miles per hour, or 300 miles per hour, will, for most cars, fuel efficiency, miles per gallon, be maximized? Well, about 47 miles per hour.

Now consider a reproducing organism such as *E. coli* eating, say, glucose. At what rate of glucose eating will it maximize biomass production per unit glucose? My colleagues and I analyzed this (Aho et al. 2012), and in the model of *E. coli* metabolism, the maximum reaches an optimum at 6 units of glucose per minute. In short, maximizing a power efficiency, here biomass production per unit glucose, picks an *optimal* displacement from equilibrium. The model fits the data on dividing *E. coli* better than the model that they grow as fast as possible (Aho et al. 2012). Furthermore, when an *E. coli* colony is crowded, "quorum sensing" slows the growth of the bacteria in the colony, to that which hopefully maximizes biomass production per unit fuel. This can be tested.

In ecology, one considers R and K selection. R selection maximizes reproduction rate. K selection maximizes fuel efficiency and is related to the carrying capacity of the niche. This picks an optimal displacement from equilibrium if the cells are doing work cycles. Odum thought along these lines and also sought a maximum power principle in ecology (Odum 1995).

I note that we are beyond Onsager and a general theory of nonequilibrium thermodynamics, once we confront reproducing protocells that do work cycles. These are not merely Prigogine's (Nicholis and Prigogine 1977) dissipative structures such as whirlpools, Benard cells, or the Zhabotinski reaction, all well known. Life is more and different than merely dissipative structures, which, Philip Anderson pointed out to me (Anderson, personal communication), lack generalized rigidity, his concept and one needed for embodiment.

What is a minimal idea of Agency without yet invoking consciousness and free will? I think, expanding upon my *Investigations* (Kauffman 2000) and later work with Philip Clayton, a bacterial agent is a system that can do the following: it can reproduce, such as a protocell; does at least one work cycle, as do cells; can sense its world, for example by receptors on its surface for glucose and toxins, thus having a biosemiotic (Kull 2009; Peil 2014) world of the states of those receptors sensing what world it can know; is capable of "evaluating" "good for me" versus "bad for me," that is, food or poison (Peil 2014); can choose what to do; and then can act to approach food or avoid toxin. Swimming toward food or away from toxin feeds or saves the protocell, so it is a function of the Kantian whole in its world and thus is a *doing*. Instrumental ought enters the universe, if not yet moral ought.

From Part II, if this system is only classical physics, it cannot have made a different choice. Actualism is upon us. But, from Part II, and without yet evoking consciousness and responsible free will, we allow quantum mechanics and measurement is real and ontologically indeterminate, then the present could have been different, the causal closure of classical physics is broken by the quantum, and Poised Realm aspects of this protocell can have acausal consequences on the classical aspect of the cell, and it can, without yet invoking consciousness, have made a different decision. I believe this is likely in protocells, for I think quantum effects at room temperature and slow power law decoherence at the metal insulator transition are likely due to location on the x axis of the Poised Realm in the ordered or critical regions, not chaotic, while almost half of organic molecules are ordered and almost half, including a number of evolved proteins, are critical so will decohere power law slowly with long-lived quantum coherent and partially coherent behavior.

And sentience?

If Part II is right in the Triad, or more generally in linking measurement to consciousness, without or with a panpsychism, then protocells and life in general may have been sentient from the start and, via real but indeterminate measurement outcomes, able to make choices but to have done otherwise. Now agency is real; it involves sentience at some level and free will, protoresponsible at some level.

If life started with minimal sentience, the evolution of life has seen its huge elaboration into rabbits and whales and us, cognitive, emotional, intuitive, and sensate. If so, free-willed choices and sentient actions have played a role, for example in niche construction by the Bower Bird, and perhaps far greater in evolution, beyond what we normally consider.

Another Example of Self-Organization: Dynamical Criticality and Reliable Action in the Face of Noise, Classical and Quantum?

Once life emerged, single-celled bacteria evolved for some billions of years, eukaryotic cells emerged and evolved, and then multicellular organisms emerged, both asexual and sexual. In multicellular organisms, their development, or ontogeny, is the subject of developmental biology. Since the evolution of encoded protein synthesis, the capacity of genes to dynamically turn one another "on" and "off," or "up" and "down," has evolved. Ontogeny is, in part, regulated by genetic regulatory networks of unknown complexity. How they evolved and their structure and logic, their resulting "dynamical behaviors" in cells, tissues, organs, and organ systems, and ontogeny as a whole, are central topics in what is now systems biology and systems medicine.

The order of ontogeny *may* be an example of self-organization that, again, does not depend upon the specific stuff, matter and energy, and hence is not due to efficient causes. To explain this, I need to briefly introduce "random Boolean nets" (RBNs), which I happened to invent and publish in 1969 and my first two books, *Origins of Order* and *At Home in the Universe* (Kauffman 1969, 1993, 1995). The issue at the time was how cells in multi-cellular organisms could be different cell types expressing different genes, when all cells had the same set of genes. Jacob and Monod in 1961 solved the problem by showing in *E. coli* that one gene could make a protein that bound next to another gene at a "cis" site and repressed transcription of the adjacent gene into its messenger RNA, hence repressing translation of that RNA into its protein. In 1963 Jacob and Monod published a seminal paper with a model in which two genes, A and B, repressed each other. Then this little genetic circuit had two dynamical steady states: A On and B Off versus A Off and B On! So this little circuit is "bistable," able to show two different patterns of stable gene activity, called "attractors," despite the cell having the same set of genes. The problem of differentiation was, in principle, solved (Jacob and Monod 1963).

Thus, it was clear cells and organisms have some form of genetic regulatory network. I hoped at the time that some class, or ensemble, of classical physics genetic networks would behave with sufficient order that it would not be too hard for natural selection to find useful regulatory network to support normal ontogeny from the fertilized egg to the newborn offspring, in us, from one to about 300 cell types by histological criteria.

To explore this hope, I needed to define different ensembles. Knowing logic but not differential equations, I thought of genetic networks to be modeled as each gene merely an "abstract gene" on (1) or off (0), a discrete state subset of classical physics. Each abstract or "formal" gene would have inputs from some genes also 1 or 0 in activity and that gene would turn off at the next discrete time moment, t to $t + 1$, if its inputs have specific values at time t. An example is the OR Boolean or logical function. If genes A and B are inputs to gene C and C is governed by the OR function on A and B, then C will turn on at $t + 1$ if either A, or B, or both are 1 at time t. For k inputs there is a finite number of Boolean functions: 2 raised to the 2 raised to the Kth power. So for $k = 2$ there are 16 logical or Boolean functions.

Thus, the ensembles I wished to study could vary n, the number of binary on-off genes, and k, the number of inputs per gene. To study such ensembles, the sensible thing to do was to *sample from each ensemble at random* to find the typical or "generic" behavior of ensemble members. I did not realize it, but this was a new kind of statistical mechanics, averaging of an ensemble of systems to find typical behaviors, rather than, in statistical mechanics, averaging over the states of one system, a liter box of gas, to do statistical mechanics, as Boltzmann had famously done.

So a random Boolean net is constructed by picking n genes, assigning to each gene at random among the n its k inputs, fixed once and for all. And I assigned at random one of the possible Boolean functions on k inputs to that gene, fixed once and for all. Having done so for all n genes, one had a random member of this nk ensemble, which has for large n and even small k a vast number of members.

One property of such networks is that each combination of activities of the n genes is a "state," and if gene updating is synchronous, the system flows along a sequence of states, called a trajectory, to a repeating cycle of states, called an "attractor." I have believed, borrowing from Jacob and Monod's little bistable attractor circuit, that cell types are attractors.

It has turned out, using randomly chosen Boolean functions, that by varying k from below 2.0 to 2.0 to above 2.0, one passes through three regimes, ordered for k less than 2.0, critical at $k = 2.0$, and chaotic for k above 2.0.

In addition to k, we consider another parameter, p, the number of 1 values in the Boolean rule, where three of the four values in the OR rule are 1. In the kp parameter plane, criticality is a one-dimensional *line*, so very rare in the nk ensemble of systems (Kauffman 1993, 1995).

Cell Types as Attractors

The sizes, or numbers of states, on state cycle attractors, vary dramatically from order to criticality, to chaos. I was thrilled at age twenty-five, and still am, to discover that for critical nets, with $k = 2$ and random Boolean functions, the lengths of state cycles was square root n! The number of states for n variables is 2 raised to the nth power. For the then thought 100,000 genes, that would be 1 raised to the 100,000th power. There are only 10 to the 80th particles in the universe. The square root of 100,000 is about 316. So critical networks, now a theorem, localize their dynamical behaviors to a tiny attractor of only 316 states out of 2 raised to the 100,000th power. Here is enormous "order for free"! If it takes a minute to 10 minutes for a gene to turn on and off, it would take only 316 to 3,160 minutes for a cell to traverse its attractors, all in the order of a few to several hours, biologically fully plausible.

I proposed, following Jacob and Monod, that cell types were "attractors." Then the different cell types of an organism would be different attractors. My initial studies showed that the number of attractors scaled again as the square root of n, the number of formal genes. An organism with 100,000 genes should have about 316 cell types. This is close to the number observed for human cells on histological criteria. It later turned out my scaling was wrong using synchronous updating, but preserved if gene updating was asynchronous. Then we thought that most DNA was junk, but it now turns out that most DNA is transcribed into functional RNA or proteins.

Well? In my 1969 paper (Kauffman 1969), I plotted the number of cell types against the DNA per cell, a proxy for the number of "genes." The results, from yeast, to sponges, to hydra, to plants and animals, show that the number of cell types is quite close to the square root of the DNA (genes I hope) per cell.

What are we to make of this? First, the early results must be reconfirmed. But how can we possibly have a scaling law for the number of cell types in organisms as a function of the DNA or functional genes per cell? There is evidence that cells are critical and, as I describe later, can have evolved for good selective reasons to be critical. If so, the generic properties of these critical systems is this scaling law, and the implications include that we can understand some or many properties of cells without knowing all the details, as "generic self-organizing properties" of the "critical" subensemble of model genetic regulatory networks of "formal genes." Again, these results do not depend upon the specific efficient cause chemistry or energy of the systems; they are formal cause laws of organization.

Further Behaviors of Random Boolean Nets and the Usefulness of Criticality

Chaotic systems show the discrete state analog of sensitivity to initial conditions that Poincaré found for three mutually gravitating objects. Small changes, flipping a single model gene from 1 to 0 or 0 to 1, unleash a vast avalanche in which 40 percent or so of the model genes in the network change their behaviors. In the chaotic regime, networks faced with even mild noise cannot act reliably. In the ordered regime there is little sensitivity to initial conditions, but the system forgets it past. To use a continuous system analogy, consider a state space with strongly converging trajectories and a bit of noise. Once in any such noisy small region of state space, the system no longer knows "where it came from." It has forgotten its past.

At criticality, the avalanches of change are a power law, with a slope of –1.5. This means that most perturbations die out with small changes in gene activities, and a few are larger. This allows the system to damp out most perturbations but also allows the critical system to coordinate behaviors on a large scale. For many reasons criticality seems useful. It maximizes pairwise correlations and, for us, maximizes the capacity to bind time reliably, remembering the past and acting reliably in the face of minor noise in the future.

Criticality is emerging to be of central importance. The brain is critical on strong data (Chialvo 2012). Cell regulatory networks seem to be critical on less data. The fact that brain and cells appear critical emphasizes that criticality is a formal cause law, not an efficient cause law, for it is independent of the stuff and energy.

Agents must act reliably in their world, by remembering their past and not being too disturbed by minor noise, yet able to coordinate behaviors over large scales. Criticality seems either one or the one optimal way to do this. This life in many aspects still unknown may make use of criticality.

Combining these results, critical systems have very small, organized attractors; few alternative attractors, or ways of behaving; and as critical, can remember their past and coordinate reliable behavior across the network. Criticality seems highly useful.

Aldana has shown that criticality can evolve easily if one selects on mutant networks for those that minimally alter the dynamical attractors, the model cell types, of the system. That is, if one supposes attractors are cell types, it turns out that these are minimally altered by mutations if the network is critical, so both evolve gracefully and can be obtained by selection to only modify cell types slowly; if not, evolution would be lethal.

Criticality in quantum and Poised Realm behaviors may be important. We do not know. At the critical line, recent work with the metal insulator transition shows that delocalized quantum wave functions in a medium have a multifractal time and space distribution, like critical classical random Boolean net power law distributed avalanches. It may turn out that criticality in the Poised Realm, coupled with decoherence and recoherence, and measurement of such power law–distributed wave functions allow some kinds of optimal coordination. It is too early to tell, but it is of interest to investigate.

Nothing limits the variables in these networks to genes or neurons; such networks may be general models for functional interactions within proto-cells, early cells, and contemporary cells. Tissues of interacting cells may be both "functionally autocatalytic" and critical; all awaits investigation.

Finally again, from Part II, we cannot exclude consciousness and responsi-ble free will even from the agency of *E. coli* or protocells or skin cells.

This discussion of ontogeny has concerned only cell types as attractors, cell differentiation as transitions among attractors, and criticality. But ontog-eny includes morphogenesis, which is a vast topic I have ignored. Here much work has been done, some of it based on the superb model of morphogenesis by Alan Turing (1953), which showed the first model of dissipative structures (Nicholis and Prigogine 1977), in that the chemical reaction system of two molecular species, excitator and inhibitor, spontaneously formed standing waves, which can be the basis of stripes on a zebra, spots on a tiger, and possi-bly sequential bone formation in limbs and other areas. The topic is central to developmental biology, yet beyond the scope of our current discussion.

Conclusions

I have discussed the possible roles of formal cause self-organization and of effi-cient causes in current theories of the origin of life, largely based on classical physics and chemistry. However, the theory of collectively autocatalytic sets is a formal cause theory of objects, transformations among them, mediated by the objects, and independent of specific efficient causes. Self-organization may explain the emergence of life, the alternative theories may explain it, or we may find other approaches. None can know what the true historical origin may have been. Its records may be forever lost. But we can hope to create life anew, and we may be reasonably close. Such life can then evolve in ways unknowable, as discussed in Chapter 4. Agency, without invoking consciousness and responsible free will, which was discussed in Part II, can be had by uniting molecular reproduction in protocells able to do work cycles, to sense their worlds, evaluate "good or bad," and act on those evaluations. If we include quantum coherent and Poised Realm behavior and measurement is real and ontologically indeterminate, then the present can have been differ-ent, and again without invoking consciousness and responsible free will, the protocell could have done otherwise. And per Part II, it may be that life from the outset partook of consciousness and responsible free will and via nonlo-cality to coordinate in ways we do not yet know. The evolution of conscious-ness, cognitive, emotional, intuitive, and sensate, is already a major subject in evolutionary biology; it may well play a role in niche construction; and it must be understood to understand the evolution of the human mind-body in its world.

I have also discussed ontogeny and the potential roles of dynamical criticality, an example of formal law properties independent of the stuff and matter. Brain and cells appear critical, for good selective reasons. We do not know the real evolved structure and dynamics of genetic regulatory networks or how they evolved, at present.

16

Knowing and Being in the World, Aspects of Our Humanity

THE LAST CHAPTER hoped to ask "what is life?" and whence came it, with a range of possible answers, ranging from life as a classical physics machine, to life in the quantum, Poised and classical realms with sentience and free-willed choice, as well as what is ontogeny. Formal cause laws were discussed and self-organization of life and aspects of ontogeny. Part II hopes to have supported our subjective pole. If so, we have our minds back, but of course, most of us never doubted our human conscious experiences and, sometimes, responsible free will.

Because "how we are human" is such a vast topic, I can only hope to point in what remains in Part III. I hope what I shall say may be useful, given thousands of years by thousands of people thinking, speaking, and writing on these topics. If we are rather lost in modernity, as I think we partially are, finding old and new aspects of our humanity that we would wish a gently transforming civilization, or multiple woven civilizations to abet, is our task.

The broad new issue I want to address arises from Part I: we often cannot know even what *can* happen, so reason cannot guide us sufficiently for living our lives forward, as Kierkegaard said.

Somehow, our set of ideas about how we know and *are* and act in the world seems in need of rethinking and reweaving. We emphasize knowing, being, and doing, but we do not unite them well. But must that be the case?

Knowing

Descartes' work is both taken as an example and taken as oversimple. Res cogitans and res extensa was his seventeenth-century attempt to maintain the subjective and objective poles in the birth of modern science. Res cogitans and "I think therefore I am," coming from "I cannot doubt that I am doubting," led to a view of mind only as witnessing and knowing the world, not as much emphasizing being, or acting in it as living humans, although he discussed these topics. Oversimplified, Descartes is a kind of idealized mind in a vat seeing the world. Much of the tradition of British empiricism, as I have noted, is an attempt to understand how such a merely witnessing mind can know the world reliably, from Hume, to Russell (1956), to Wittgenstein's early *Tractatus* (1921), to the logical positivism (Suppe 1999), of my philosophic youth which proudly claimed "only those statements which are empirically verifiable are meaningful!" I always loved this in my young Oxford philosophy days, for the very founding statement of logical positivism just quoted is *not* itself empirically verifiable. How did they miss that rather salient fact? Finally, via the later Wittgenstein, *Philosophic Investigations* ([1953] 2001), and language games that cannot be reduced to lower language games, and the failure of sense data statements with Russell (1956) and the *Tractatus* (Wittgenstein 1921), there is no basement language of propositions with which to know the world. With this I fully agree. And, if so, no entailing theory of everything will entail legal language or the practice of law. (Ah, philosophy. I do love it!)

Borrowing from my cousin, Richard Melmon, I raised the issue that language probably started as metaphoric, or even gestural, not propositional. Let's focus on this again. Metaphors, such as "Juliet is the sun," are neither true nor false, but they can powerfully elicit responses that can orient our lives. Metaphors, here, can evoke new uses of the screwdriver from us. With them we sometimes solve the frame problem of computer science that cannot be solved propositionally. Perhaps we use metaphors to solve the frame problem via a unity of consciousness due to quantum entanglement and solve James' problem of combinations, solved if there is a single wave function for the entangled particles.

But metaphors really are not true or false. Then we invented propositions, "The cat is on the mat," obeying the law of the excluded middle at the huge price of categorizing the world into cats and mats. And then syllogisms, as noted: "All men are mortals, Socrates is a man, therefore Socrates is a mortal." With this we get logic and entailment and later with numbers, equations and physics with mathematical equations.

I've noted, and think we will agree as we go forth, that no prestated set of propositions can exhaust the meanings of a metaphor. Yet we use metaphors, hence beyond any prestated set of propositions, for example to find new uses of screwdrivers. Furthermore, I repeat with minor pleasure my claim that *if* no prestated set of propositions can exhaust the meanings of a metaphor, and if mathematics and logical entailment require propositions, then no mathematics can *prove* that no prestated set of propositions can exhaust the meanings of a metaphor. I love this little argument and its negative result, I confess.

But we "know the world" with metaphors. So we know the world without that knowing being true or false, and we live, feel, evaluate, judge, and are guided to action and are in the world with metaphors.

One extreme case of propositions resides in mathematics. We do love mathematics, with its astonishing wonder from Euclid to Einstein. Before non-Euclidian geometry was discovered, we thought that mathematical truths described the very world, the Pythagorean dream. With non-Euclidian geometry we gave that up. We reduced our dream to that in which David Hilbert in the nineteenth century hoped to axiomize mathematics and show that it could be complete in the sense that all true statements given the axioms were derivable as theorems from the axioms. "No," said Gödel, in his famous proof of "formal undecidability," either the theorem is formally undecidable from the axiom set, or otherwise, the axiom set leads to inconsistencies. We cannot even be sure we have a consistent axiom set, and often we do not. There is no basement language to mathematics either for even modestly rich axiom sets. The cousin of Gödel is Turing, who showed that no Turing machine could formally decide if another Turing machine program would halt (Turing 1968). Such Turing machine behavior is "semidecidable." It is deterministic, but we cannot say before running the program what it will produce. Perhaps, even in classical physics, much of reality is semidecidable. G. Chaitin (Chaitin et al. 2012) invented "omega," the probability a computer program will halt. Each binary digit of this decimal, say, 0.1010001, is not derivable, but a "crystal of pure creativity," to use Chaitin's magical metaphoric phrase. There is, here, no mathematical basement language to derive omega (Chaitin et al. 2012).

We hope some mathematics does describe the world, general relativity, quantum mechanics, and the standard model at present, for example, and perhaps they do, but we are forewarned by Wittgenstein ([1953] 2001) and legal language. We cannot "know" everything in some deductive way from a final theory, such as my legal status at trial, and we cannot know ahead of time the evolution of the biosphere. Propositional reason fails us—no basement propositional language with which to know, no knowing all that *can* happen. But life goes on.

Being alive in the world is more important than knowing. Until a protocell existed in the universe nonergodic above the level of atoms and could evolve, there was no selective advantage to sensing the world—hence biosemiotics, knowing the world. Being alive precedes knowing, and knowing is only part, along with "hedonic" without or with sentience evaluation "good or bad for me" and doing, with our being alive and getting on with it in the world. Being, knowing, evaluating, and doing have always woven together in life, without or hopefully with sentience built from a panpsychism. Earliest life may have known and been able to choose with some form of free will, but it could have chosen otherwise, so the present could have been, counterfactually, different, and life is not a classical physics machine. We do not know this yet, but given quantum biology (Engel et al. 2007) and the poised realm (Kauffman et al. 2014), I'm betting on it. Life had mind from the start, I say, but cannot yet know. Perhaps one day we will be able to know by finding quantum and Poised Realm behaviors in simple life, or new life we create, finding couplings among organisms by quantum nonlocality and possible measures, some cousin of Radin's still early results where human consciousness appears to alter measurement (Radin et al. 2012, 2013), that *E. coli* can also alter measurement. If we can ask humans if they are paying attention or not, can we "ask" *E. coli*? Perhaps by their diversion and operant conditioning? Who knows?

If life started as sentient, Mind has evolved along with life in the evolving biosphere, and it has presumably played a role in evolution, to which I return later.

Art and Science in Our Humanity, Forms of Knowing and Doing

A major article, "The Two Cultures," by C. P. Snow ([1959] 2001) decried the split between art and science. It was published in the 1970s. The split continues, then with art on top, now with science.

I hope to address this issue, expanding it from merely "knowing" to knowing, feeling, sensing, intuiting, being in the world and doing, by returning to Quine's "Two Dogmas of Empiricism" (1953). In the second half he argues that no hypothesis confronts the world alone, but in a web of statements (propositional) of the facts and a web of interwoven hypotheses. Given disconfirming evidence, we can choose to save what we wish, such that the total web fits our awareness of the world "at the edges." Here Quine is concerned only with "knowing" propositionally. But in the first part of his article, Quine doubts the clarity with which words or propositions "mean" and specifically pick out aspects of the world. Consider "ostensive definitions." I point to a rabbit and say "rabbit" to you. Am I pointing to the rabbit, the white hair of the animal, its ears, the distance from the rabbit to the nearby tree? Ostensive definitions are metaphoric, and neither true nor false. But if so, all our knowing is metaphoric at root. There is, with Wittgenstein (1953), no basement language at all, even for mathematics and physics.

On Quine, then, even propositions are, at root, metaphoric, and art is metaphoric. Does this not give grounds, not to "unite" science and art," but to see that they are both our ways of using words, music, pictures, to be in the world, know in the world, feel in the world, sense in the world, evaluate in the world, create in the world, and act in the world, scientific, artistic, mathematical, practical? It is just us making our alive ways of being and doing in the world together as we co-create it. Amoebae probably make their ways in the world too, knowing as they go.

This again becomes Wittgenstein's language games, ways of being in the world with words and actions. With Wittgenstein, we begin to span from knowing to being in the world and doing: I am found guilty of murder and hung, aspects of the world not derivable from any theory of everything, for there is no basement language with which we know, are, and act in the world. Wittgenstein speaks of language games as ways of being in the world. He was right. So are gestures, body language, kisses, touches, and running from tigers without a word. So is writing poetry.

If Quine speaks of a web of hypotheses and statements of fact that web the world we wish to know scientifically, he spins a wider tale of language, metaphoric and propositional, with which we are, know, and do in the world. We need a wider conceptual web that spans all of this, including first-person phenomenological (Merleau-Ponty 1969), accounts of embodied experience, intention, motive, and action. A naive theory of psychology says that "Sometimes we do what we say we will do for the reasons we say we will do them." Well, sometimes that too is true, the folk psychology derided by the academy but upon which we base our lives. Read Shakespeare again: "Tomorrow and tomorrow and tomorrow. . . ." So? Do we really want our conceptual web to span being, sensing, knowing, feeling, evaluating, and doing to not be rewoven into a life-affirming web, embracing our subjective pole fully? Otherwise "we have met the enemy and it is us." Rather, we are fractured into third-person objective knowledge, scientia, and first-person living, woven uncomfortably together in our scientistic modernity.

Superb minds have been here, beyond human "knowing." A strand in Western philosophy runs from Kant through Hegel and Schopenhauer to

Nietzsche (Nietzsche 1968), and more recently Habermas (1971), asking how we are humans in the becoming history of the world. Kant sought transcendental foundations, then denied by Hegel, who sought instead historical becoming with his thesis, antithesis, and synthesis, but he thought there would be an end to emergent history. I so doubt this thought, also explicit in Marx; history in the antientropic universe above the level of atoms seems almost surely to be a persistent, unprestatable status nascendi. Merleau-Ponty and others struggle to found an embodied phenomenology of humans. What is it like to be *E. coli*? Stentor? What is it like to be a bat (Nagel 1974)? Mind and phenomenology, along with life, almost surely evolved. Vedic philosophy has been here for thousands of years. And science in the West dares to ignore a thousand-year tradition exploring our subjective pole? Were these peoples simply wrong?

How Are We in the World? Intuition Beyond Knowing and Doing

Since ancient times, we are familiar with reason or rational thinking, feeling, and sensing the world, knowing and doing and being. Jung (1971) included these three but added *intuition*. What in the world is intuition? Always the "Muse." When chemist Kekule, thinking about the structure of the benzene molecule, looked into the fire and the image of a snake biting its tail flashes into his mind and he "realizes" that the benzene molecule might be a *ring*, what happened? When the man in Tokyo with the iPad in his crowded apartment thinks to scan his book into his iPad and then sell his book, he "*realizes*" his new business opportunity: to scan other people's books into *their* iPads, sell the books, and take 5 percent of the sales as his profit. What has he done? He has seen an opportunity, realized its portent, and grasped the opportunity to start a new and successful business. What did he do?

What did Einstein do when he imagined riding on a light beam and invented special relativity, not reducible to Newton, and when he realized that if one stood on a weighing scale in an elevator in free space accelerating at 32 feet per second per second, the registered weight would equal that if one stood on a scale on Earth, so accelerating motion was related to gravity, inertial mass was equal to gravitational mass, and he invented general relativity? Einstein did *not* recombine old ideas; he invented new questions, crystals of pure creativity, to ask of nature! Whence? And he invented new concepts to reframe the world. No basement language in science, no basement language, gestural, metaphoric, propositional, in living our lives. We can and do see and find new questions, issues, opportunities all the time, not hinted at before.

What is it to see a new opportunity? We do it all the time. What is it to see an opportunity no one else has seen, hidden in the Actuals about us? In science, we think our business, given a prestated problem, is to propose hypotheses and then test them, either to confirm or deny. But this book is full of me doing something else; I am but one case of one living being finding new questions of nature. Only after the question is asked can a hypothesis about an answer even begin to be formulated. Asking a new question of nature is like seeing a new business opportunity. How do we do it?

Some think that intuition is merely imagination of recombinations of old things, the body of a horse with the head of a man. But the realization that the rigid engine block could be used in a new way, as the chassis, to invent the tractor, is not such a recombination. We are doing something else, something we know, screwdriver-like, that is not propositionally algorithmic.

This new seeing creates new Actuals, and beings and doings. The iPad business, the tractor, and tests of general relativity, in turn, elicit new unprestatable ideas, acts, and becomings of our human world.

In short, with Jung (1971), I too think intuition is a fundamental irrational aspect of human mind, and actings on it are part of our human creativity in a creative universe.

I think the human mind-body system is as in Part II. And I think mind is quantum coherent, Poised Realm, partially quantum coherent or partially decoherent and recoherent, breaking the stalemate of the causal closure of classical physics to answer Descartes and partially "classical." And I think that quantum superpositions are real, do *not* obey Aristotle's law of the excluded middle, and so bespeak an ontologically real Possible, res potentia. Metaphors do not obey Aristotle's law of the excluded middle. There is evidence from Liane Gabor and Detrick Aerts (Gabora and Aerts 2002) that human categories have quantum, not classical logic. The evidence is weak, but it can be improved or disproved.

What if mind is partly quantum coherent in our mind-brain system and if that coherence is, in fact, res potentia, ontologically real Possibilities? Perhaps the unconscious mind is quantum coherent and measurements happen with flashes of consciousness and that is our access to the Possible and constitutes intuition. And if n entangled particles are involved, they are described as a whole by a single wave function, they are not "independent particles," so when measured we may get a "new qualia," not a mere recombination in our imagination of the body of a horse and head of a man.

Perhaps Kekule gets the snake biting its tail. Entanglement may, as noted, solve James's problem of new wholes from combinations of atoms of consciousness (James 1909), and we solve the frame problem, but not propositionally. We find new uses of screwdrivers intuitively via entanglement and measurement of new wholes.

The Emotional Self-Regulatory System

Katherine Peil has published a recent paper on emotion in a broad sense (Peil 2014). Think, she says, of *E. coli*, with receptors for glucose and pH, and toxins and so forth. Without or with invoking consciousness, perhaps with "sentience" that leaves open the issue of conscious or unconscious, the states and rates of change of states of these receptors are the "world" the *E. coli* "knows." This biosemiosis is real (Kull 2009). Then, argues Peil, the cell must "hedonically evaluate" "good for me?" versus "bad for me?," which she wishes to give a hedonic valence, the root of emotion which may be one of the first senses, if *E. coli* is proto-conscious, as we are aware of pain. After evaluation, act, and act reliably if possible. Peil considers this total integrated system: sense the world, evaluate it hedonically, and act, to be a "self-regulatory" system. A wonderful feature of *E. coli* is that it uses both phosphorylation to start flagellar counterclockwise rotation from the rotor motor, and more slowly uses methylation of sites to *record* that "glucose" is this way. These records seem to require a classical-enough world to be stable. A remarkable feature of these records is that *E. coli* will swim toward where the glucose *was* but no longer is. Given a record, *E. coli* can be "mistaken." Then with more evolution, as Dennett says in *Darwin's Dangerous Idea*, later we can run the record motor without acting and let our records die (1995). But this is true, only, I claim, if we and *E. coli* could, contrary to fact, have chosen to do or act otherwise. If not, if we and *E. coli* are classical physical zombies, records or not, we, zombie-like, are triggered to "act" or not, as evolved machines, and live and die by our "action" for selection's sifting. In that case, we cannot let our hypotheses die in our stead.

Peil (2014) broadens this to four negative fundamental emotions, fear, anger, disgust, and sadness, and a positive emotion of joy in us humans. From these she thinks complex emotions like admiration, gratitude, schadenfreude, empathy, love, and others emerge. Her views seem right. And also the psyche is very rich in its conscious and unconscious imagery, beliefs, feelings, and intuitions that undergird our psychic life, including emotions, which, she rightly says, constitute the fundamental source of values, choices, and doings.

If panpsychism is right, life evolved with pressure, temperature, consciousness, and will from the start, elaborating it in myriad ways, being, sensing, knowing, feeling, evaluating, intuiting, and doing.

If life evolved mind from the start, there may be signs of it. Piel has looked at the data on Stentor (Bray 2009). This ciliate, a single-celled organism, lives attached to a substrate and is a filter feeder. It has a "mouth," stomach, stalk, and foot. Given a toxic chemical stimulus, Stentor rotates its mouth away from the toxin, as if in pain; it then empties its stomach as if vomiting in disgust; and it then curls over as if avoiding the toxic stimulus. If the stimulus continues, Stentor uproots itself and uses its stomach to crawl away. These, say Peil, seem as if they may be signs of the same fundamental negative emotions in an early evolved single-celled form of life. Can we test this now? No. But if we could ever accept that human conscious experience modified measurement, we would have a "consciousness meter" and might test if Stentor is conscious! Maybe. If so, what of eubacteria, archibacteria, Anabena, Hydra, Physarum, worms, fish, amphibians, reptiles, birds, and mammals? What of ferns and grass and plants? Perhaps we should at least suppose we might one day examine the evolution of mind up to us. Darwin studied the expression of emotion in man and animals, published in 1872.

If mind evolved from the start, what has been its role in evolution? We understand niche construction, from Bower Bird, to us and our economy. But how widely has our conscious and free-willed behavior played a role in the evolution of the biosphere? We have ethology, which studies behavior and guesses phenomenology. How can we study this in a Quine-like widened worldview with a full subjective pole as an integral part? How can we *be* in it, cat and horse friends, we loving our pets and they us, sensing our joy and sorrow with licks and pawprints. Well, we *are* in this, pets and walks in the woods to marvel our birthdays away.

Magic and re-enchantment are emerging beyond law in the evolving biosphere.

Institutions and Our Beings, Knowings, Feeling, Intuitions, and Doings

No man is an island. Our lives evolved as social primates, the last major biological evolutions largely in the Paleozoic, but some undoubtedly now. Our fulfillment and grief come not only as individuals and families but have evolved from the clan to huge societies, with institutions of many types on all scales. This is the stuff of anthropology, sociology, human history, the ongoing world of business and the practice of law and plumbing. Among these social forms are institutions, from the Yankees to the banks too big to fail. We often identify our own aims and emotional lives with the aims of our organizations, and we find grief or fulfillment in them. The Yankee coach whose team won the Pennant is delighted. But institutions, like organisms, achieve functional closure or sufficiency in their worlds, and like organisms, they adapt in unprestatable ways into the Adjacent Possibles we all, often unwittingly, co-create. I return to this in the next, final, chapter, for these also give rise to the power structures that strangle us. It seems of high interest to try to begin to map from the functional sufficiency of each cell in a multispecies microbial community, where collaboration and competition in myriad ways must be happening, to our economy with similar features, to our social organizations in general. How does microbial life evolve functionally into the Adjacent Possible? How do we and our social organizations evolve? The huge question is how do we do this wisely? I return to this in the next chapter.

Whence Moral Ought?

How we have struggled over this issue, because for given any claim to moral ought, we can ask, but is this moral ought really moral? On what ground? God? Kant's noumenal locus of "moral ought"? Utilitarianism, "the greatest good for the greatest number," is unable to solve the problem of how justly to distribute the total "good." And morality and "Justice" face further the fact that we do not always know the consequences. Rawls's (1971) post-Kantian efforts at Justice with his two rules: We behind the veil of ignorance in face of a society's institutions and would we choose them not knowing our station? And his effort at distribution: "The most possible to the least of us." But this is inadequate; our institutions evolve in ways we cannot say (Devins et al. 2015), including the US Constitution where the Commerce Clause has been used for the drug war that has filled our prisons with too many black youths. We cannot design institutions; they grow, partly helter-skelter, with unintended consequences, as discussed in Chapter 5. No one time bargain, à la Rawls, will suffice. Our institutions, and we living in and through them, evolve in unprestatable ways.

Yet at least liberal democracy seems plausible since Darwin, Smith, and Locke, as argued by Arnhart (2014). We are social primates, and it seems likely that much of our morality evolved in the Paleolithic. Yet we are not frozen with that morality. The Hammurabi code of "an eye for an eye" was a huge advance on ten eyes for an eye. The Ten Commandments were an advance. So was the evolution of English common law. A wonderful case in point occurred in England from 1770 to 1830. At the start, if you stole my sack of gold, and were found guilty, you were severely punished. The violence in the crime did not matter. Over the next sixty years, reference to violence in the crime in court records increases and, rather wonderfully, because under common law, juries cannot be held further responsible or punished, for their verdicts, juries began finding innocent those who were clearly guilty. Why? Because society came to feel that the punishment was too severe to be "fair." This is called "jury nullification"; it was not a judge ruling precedent, but it is practiced. In much of the United States, I'm told, it is a felony for a judge to tell the jury that it can nullify. But given the current debate about marijuana, legal in some states, in 2014, but not nationally, a major US Justice official said publicly recently that punishment for small amounts of marijuana would be defeated by nullification.

Are our ethics settled? Rawls likes the notion of "reflective equilibrium." But in the fourteenth century, the English practiced hanging, and drawing and quartering. Would we now? What is reflective "equilibrium"? I doubt we ever come to reflective equilibrium. Our ethics have evolved since we were hunter-gatherers in roughly egalitarian clans, as Locke learned, to agriculture and the onset of massive civilizations and hierarchy and domination, and the code of Hammurabi, to the Magna Carta, to, as Arnhart argues, something like a Lockean republican democracy with a hoped-for balance between my freedom and the collective use of commons with the free rider issues it raises and the drive for power.

But has it worked? Not really. In the United States, the balance of forces, executive, legislative, and judiciary, is tilted against the judiciary, and the power of the executive branch has grown massively to include the US NSA spying on all of us, even the Prime Minister of Germany. Our Constitution has evolved, neither as the "originalists" might have wished, or the "living document" proponents might have wished, but in myriad unintended ways (Devins et al. 2015).

We think we know, from Newton, Darwin, Smith, and Locke, how to wisely govern ourselves. Do we? We have ground for serious doubt, and an examination of more bottom-up enablement and multiscale coordination, and less belief in top-down control by experts who are not experts is needed. We cannot know!

Improvisational Comedy Again: More Than a Metaphor for Real Life

I've discussed improvisational comedy. I think it is more than a metaphor for our lives. Again the rule is you must accept my line and build on it in a comedically appropriate but unprestatable way. "Here is a silver platter with a steaming pile of horse crap," I say. *"Nope"* is not good. "Wow, where's my pie plate made of dried dung?" you say. If we go around the four of us, each line does not cause, but enables a comedically appropriate adjacent possible but unprestatable set of possibilities for your next line. We go around the circle and end up, if any good, with a skit none of us could have foretold, but it is funny. We jointly co-created the skit, not knowing what we would do. So is it for improvisational jazz groups or even a single violinist improvising while playing Mozart. We co-create our worlds not knowing. If you have never heard "Beyond the Fringe," Google it. It is the best of British humor in the past 100 years. About the actual Great Train Robbery, the head of the Scotland Yard explains on BBC, "Not a robbery of a train at all! No train disappeared! Last one was 1878, found in a week on a siding in Wales. Large bulky objects, difficult to conceal!" "Do you suspect anyone?" asks the BBC man. "Oh yes, we suspect thieves!"

How do we *do* this?

Life and human life is not merely preplanned, although it partially is, and these are partially the roles and rules that are enabling constraints, like grammar, that enable us to get on with getting on. But that getting on with it, British stumbling through, *is* improvisational *life* becoming into the possibilities its becoming enables.

Poetry and Art Again

Science, as Snow noted, is at poor peace with our full humanity. To wit, Shakespeare, who dominates writing plays as no other artist dominates any other art. It is said normal humans have a theory of the mind of the other to six levels. Then Shakespeare reached the seventh.

Poetry bounced off the walls of science, as I noted, from Pope, "God said let there be Newton, And all was light," to Keats, "Science with its rule and line," robbing us of our humanity. I so hope Parts I and II may help restore some of that humanity to us and help in some small way to unleash us.

Poetry and plays were so central to Athens 2,500 years ago. Now our vestiges are movies. And somehow we got from *High Noon*, with Gary Cooper against the wishes of his Quaker wife, finally taking up his Colt to blow away the four bad men at noon and then putting his gun down forever, to *The Terminator*, not even human. How did American culture achieve this? Yes, there is much more wisdom in common culture than I speak of here. But are we not rather debased? A fine novelist told me some time ago that in the first half of the twentieth century America produced superb art and literature. No more. We are, he said, a civilization with little culture. Reminds me of the billboard in a Wisconsin town: "All of the yogurt, none of the culture." Do I overstate? Maybe, in this, our app-driven modernity, we are reduced, as Gordon Brown as Prime Minister of the United Kingdom said in Strasbourg, to price tags.

"Out, out, brief candle! Life's but a walking shadow, a poor player that struts and frets his hour upon the stage and is heard no more. It is a tale told by an idiot, full of sound and fury, signifying nothing." How do we write this? This *is* our humanity. And consider this lovely excerpt from Dylan Thomas's "Poem in October":

> It was my thirtieth year to heaven
> Woke to my hearing from harbor and neighbor wood
> And the mussel pooled and the heron
> Priested shore
> The morning beckon
> With water pray and call of seagull and rook
> And the knock of sailing boats on the net webbed wall
> Myself to set foot
> That second
> In the still sleeping town and set forth.

And then there is Pavarotti singing "Nessun dorma."

This is Darwin's tangled bank, not of the biosphere alone, but our humanity, alive, singing together, having touched the finger of God. Away then, let us away together. But "how together?" We kill, yes? Then: Whither? To be fully alive. But what is that?

What is it for humans woven together with one another and nature and all our diverse cultures have become and are becoming, to be fully alive? This, I believe, is "whither civilization," if we can begin to speak it so. Whither a new Axial Age?

Conclusion

I have sought in this chapter, treading where the Axial Age trod 2,500 years ago, to touch some aspects of life, living, doing, feeling, sensing, intuiting, knowing, and being in the world as alive humans alone and together. We need a new conceptual framework that allows us to speak of life in interwoven language and metaphor, recognizing the richness that, in fact, we know, feel, sense, intuit, act. We will not derive all this from a Final Theory, not if a jury can hang me for a crime that cannot be expressed in physical language but by a legal system that itself evolves in unprestatable ways.

No one of us can or will do the weaving, nor will it ever end.

17

Beyond Modernity?

CAN WE PARTIALLY CO-CREATE A WOVEN GLOBAL CIVILIZATION THAT SERVES OUR HUMANITY?

IF WE ARE, by Parts I and II, partially set free and have by these efforts more of a basis for real scientific hopes for our full subjective pole, perhaps we can now a bit more easily have our full humanity, its better, its worse. "What can we do with the crooked timber of mankind?" wrote Immanuel Kant. Crooked we can surely be, best face it and seek the conditions that foster our better selves.

Max Weber said, "With Newton we became disenchanted and entered modernity." Yes, and to recount: Newton leads to science leads to the Enlightenment saying, "down with the clerics up with science for the ever betterment of mankind." This leads to the Industrial Revolution, then to the modern world, and then to whatever postmodernity or post postmodernity may be.

We remain disenchanted. There is *no magic* in the entailed world of Newton, leaving the Romantic poets to cry out for that magic and a humanity lost: "Science with its Rule and Line," wrote Keats.

As Andreas Weber (Weber 2013) and Larry Arnhart (2014) rightly write, Adam Smith and Darwin are, with Newton and Locke, at the core of modernity's mythic structure. Adam Smith of the Scottish Enlightenment gave us the shoemaker and candlemaker, each acting for his or her selfish interests, yet leading as if by an "invisible hand" to the welfare of all. Never mind that Smith also wrote on the "moral sentiments" in another book that is ignored. Capitalism roots itself in Smith, who founded modern economics. With Riccardo, and the advantages of nations, we have a massive preference for "free markets." Never mind that the US colonies created the foundations of what would be the early US economy behind tariff walls and by copying locally British imports. Free trade? Just ask the World Bank and the International Monetary Fund, still wedded to these ideas and the Washington Consensus as poverty flourishes. Maybe we sometimes need tariff barriers in very poor countries behind which they can build a self-sustaining web of complements and substitutes at their *own* technological levels that generate an internal market, not wedded to global prices.

As Arnhart (2014) says, Smith is one strand that says that society can be self-organizing from within, needing no external authority for its structure such as the medieval Church view of an outside source, God, for the well-run structure of society. Then, again as Arnhart writes, Darwin gives us for the first time in history, evolution by natural selection, hence the appearance of outside design, with no designer. There is no Watchmaker, the response to Bishop Paley's appeal to the adapted organisms of the biosphere as watches that demands a God as designer. But the narrow reading of Darwin is "survival of the fittest," individual selection for those with more offspring, in a competition for limited resources, an echo of Smith's selfishness yet for the benefit of all, leading, it would seem, to Darwin's tangled bank teeming with life.

Four superb minds, Newton, Darwin, Smith, and Locke, form the core of modernity's mythic structure. The world, from Newton, is entailed, a machine that can be known and substantially controlled. We need only be scientific and rational and use that rationality well. We will know; we will master. We will as we unleash thousands of chemicals into the atmosphere with no idea of their effects, overfish the Grand Banks of Newfoundland, and cannot reestablish the fish-rich seas. We will master? Only the Superman of my young boy's comic books, in the absence of kryptonite. Smith and Darwin yield society organized from inner wellsprings. In the simplest view it is the selfishness of the shoemaker and candlestick maker, each acting for his or her own benefit, that enriches society. And it is true in many respects. There is a three-day supply of food in most major cities. The invisible hand of the market fills the markets. (But worries about the fragility of the global supply chain are growing and should. We cannot just design it when we do not know what can happen. We must prepare for the unknown and both "jury rigging" as unforeseeable solutions, and the implied incompleteness of contracts are issues we should face.) From Darwin, natural selection springs from Malthus, who wrote that exponential growth in population and linear growth in food supply meant eventual death. "At last I got hold of a theory to work with," wrote Darwin in the margins of his notebook. So natural selection, in the first image, is competition for scarce resources, and the better adapted will be selected. At root, this image too is "selfish," for each organism adapts, thanks to heritable variation and natural selection, to better fit its environment as measured by the mean number of its offspring.

Newton, Darwin, Smith, and Locke are the intellectual pillars of modernity and our mythic structure. Locke studied Hobbes and Rousseau. For Hobbes, man is selfish and brutish, and bands in a social contract into society for protection. For Rousseau, man in the state of nature, is cooperative. The New World was being "discovered" and relatively simple societies could be studied and were. A cooperative egalitarianism was found in many of these societies that were not dominated by the Incas and Aztecs. Locke studied these well, in reports from the New World. But after the agricultural revolution allowed massive accumulation of stored wealth, massive power hierarchies arose in Egypt and Mesopotamia. Millennia later, in the sixteenth to seventeenth centuries of Europe, kings ruled by divine right. Locke's political philosophy became ours in most of the West: from Newton, government as a balance of forces, in the United States, three branches of government, executive, legislative, and judicial. A foundational freedom for the individual and his or her liberty and right of ownership of property, the basics needed for Smith's invisible hand. "Life, liberty and the pursuit of happiness," wrote Jefferson. The pursuit of happiness, human flourishing in a broad sense, has become, to some extent, limited to be the driver of economic trade to fulfill some of our human desires, and create wealth and profit, the wellspring motive forces of the invisible hand. "We are all reduced to price tags" wrote Gordon Brown as Prime Minister of the United Kingdom. "Wealth" is not, however, human flourishing. Arhnart (2014) writes further that Locke envisioned a republican government that balanced the egalitarian aspects of human social life with the drive for dominance evidenced in the first agrarian civilizations after agriculture allowed massive accumulation of stored wealth.

In stark form, this constitutes the mythic structure of modernity. Much of it is wonderful, and much wonder has ensued.

What is missing? First, Darwin, narrowly taken, underestimated collaboration. Consider the 150 bacterial species in your gut. You must have them to live. What are they doing for one another functionally with their myriad molecular and cellular behaviors? Most are not competitors, but commensals. We do not have any idea how these may be functionally collaborating so that each member of each species achieves a functional, not closure, but sufficiency, in its world with the other species and your cells. The evolving biosphere is not just "nature red in tooth and claw," which was not Darwin's phrase. Nature is Darwin's "tangled bank" teeming with functionally interwoven, collaborating, and competing species, an ongoing flood of becoming life with 99 percent of all species dead, but still a flood of becoming. Were we to better understand this functional interweaving, how it comes to be and continues to become, and be the ever-evolving balance of nature whose members are ever-changing, as life ever "finds a way" to enter its Adjacent Possibles, we might find lesions to apply to the ongoing historical becoming of human socially interwoven life, where we too collaborate and compete, grow, and wither in myriad ways as individuals and members of so very many overlapping organizations on all scales. Adam Smith's invisible hand alone will not, a theorem by Arrow (Arrow and Debreu 1954) shows, lead to total definable "social welfare." The problem is evident in utilitarianism: the greatest good for the greatest number. But how shall that total good be apportioned among the great number? Utilitarianism has never answered that issue, and Arrow shows there is no unique solution to it. With the broad legal notion of "equity," or fairness, and our broad ethical sense of "fairness," we struggle ever with this. Our moral sense may well have evolved partially in the Paleolithic, where fair sharing was wise. Interestingly Arnhart reports that in simple hunter-gatherer societies, meat is shared widely in the clan, and vegetables are held by each for the family. Why? One hypothesis is that finding meat is risky, so sharing widely is to the advantage of all in the clan and "fair." But finding roots and plants is not so risky, so if I have them and you do not, you did not try hard enough, so sharing is not fair and you are a free rider. The long evolution of social practices, contract law, and property disputes bears witness to our evolving sense of what is fair. But we know this, 3,000 years of the evolution of legal laws in our thirty or more civilizations, our political, ethical, and religious traditions, have struggled with our living with ourselves and together. Perhaps we need to attend more attentively to this we already know and bring these issues ever more starkly before us. As "price tags," we ignore what we know.

It may be that the evolving biosphere is dynamically critical. This could be very important. Evidence for this is that extinction events come in small and large "avalanches." If log size of avalanche is plotted on the x axis, and log number of avalanches of each size is plotted on the y axis, one obtains a straight line through the data sloping down to the right out the x axis. A straight line in a log-log plot means that one variable is the other raised to a power. This is often a signature of "critical behavior," which has just such power law avalanches (Kauffman 1995). Perhaps speciation bursts post extinction events are also a power law; we could find out. If so, then criticality of the biosphere, which is a feature of many coupled nonlinear systems (Kauffman 1995), may be hinting at something deep about how functionally coupled, cooperating, and competing organisms in the evolving biosphere co-create their small-scale and large-scale organizations, from individuals to the entire biosphere, with no one in charge. But the same may be true of the economy. Schumpeter wrote of "gales of creative destruction" (Hanel et al. 2007). Here the horse was replaced by the car, the horse went dead as a means of transport and with it, the buggy, saddler, watering trough, and horseshoes. But with the car came a demand of oil and gas, hence a new vast oil industry, paved roads, traffic lights, traffic police, traffic courts, motels, and suburbia whose inhabitants needed cars to go to the city to work and return. Recent work (Hanel et al. 2007) suggests that such gales may also be a critical power law. Perhaps our evolving, functionally coupled economy is critical as well, again across all scales and with no one in charge. While neither the becoming of the biosphere nor economy is entailed, there may yet be forms of nonentailing laws that are independent of the specific stuff and processes involved, perhaps formal cause laws, not efficient cause laws. Is criticality a clue to how multi-scale functional organization emerges in the biosphere and economy? If so, it is deeply important to understand it and what it means for us.

A second aspect of modernity is its overreliance on rationality. The Enlightenment is, after all, the Age of Reason, fruit of Newton and others: we can know, we can master, we can, substantially control using reason and science. Oh? Not as often as we think. The Peruvian government, acting to protect the Amazon forest, passes a law based on satellite images of the forest canopy, to not disturb the canopy. Locals find a loophole in the law: cut down trees shorter than the canopy and sell the timber. Pass a law and someone will find a loophole no one expected. Control by laws thought wise? How often do we have failure by typically unintended consequences? Devins et al. (2016) discuss the unintended evolution of our Constitutional Law as a powerful further example of this at the highest levels of law. We are surrounded and engulfed by unintended consequences we often cannot have foreseen. If we cannot foresee them, we cannot reason about the issues. The wonderful Arrow-Debreu theorem of competitive general equilibrium demands that we begin by prestating all dated contingent goods, for example, a bushel of wheat on your doorstep next Tuesday if it rains in Boston on the Monday before. But we cannot prestate all dated contingent goods. First, even with existing noncontingent goods, we cannot prestate all contexts, which is the frame problem. We cannot prestate all the uses, hence contexts of a screwdriver, hence markets for new uses of screwdrivers. Second, no one, when the

Turing machine was invented in 1933, foresaw the mainframe computer, personal computer, word processing, file sharing, the Web, selling on the Web, content on the Web leading to Google, and social media on the Web and governmental spying trying to gather data on customers from Google. The becoming of the economy, like that of the biosphere, is typically unprestatable and governed by no entailing law. Reason? If we cannot know what *can* happen, reason fails us. But we must live forward anyway. We must more consciously attend to the fact that we do live forward without knowing and attend to how we do so, and ask, what is wise. The Soviet Union collapsed, who knew? What if the size distribution of historical "events" somehow defined is critical? What would that tell us?

Third, the Age of Reason downplayed our emotional lives. But this is huge. Our fulfillment, our flourishing as humans alone and woven into our roles in our multifaceted societies, is driven by and mediated by our emotional lives, our only biological evolved source of value. It is not an accident that we in modernity easily remember Smith's *The Wealth of Nations*, but few know his work on the moral sentiments, or Darwin's *Origins of the Species*, but not his work on animal emotions. What is the "happiness" of which Jefferson wrote soaringly: "We hold these truths to be self-evident, that all men are created equal, that they are endowed by their Creator with certain unalienable Rights, that among these are life, liberty and the pursuit of happiness." But "happiness" is human flourishing. Down the ages we debate what is the well-lived life. For Emerson and Thoreau it was the well-cultivated life in which one gardened one's capacities (Emerson 2000; Thoreau 1854). But I argue for more: We become in ways we cannot know alone and together. If human flourishing, as Peil (2014) argues, is a weaving of our negative and positive emotions, the latter, joy in a broad sense, then it finds expression in our growth, hence: "Live the well-discovered life!"

A fourth aspect, of major importance: we miss spirituality at a fundamental level. We give our faith to science and rationality. From this and our adoration of technology, from washing machines to thousands of apps, comes the overwrought scientism that smirks at "spirituality" as unseeming, foolish, or, in the strongest case among the neo-atheists, overconfident in their science, that any belief in any form of God, monotheistic or not, is stupid. One of the most transformative moments in my life was, at age fifty-two, in 1992, to participate in a small Gehon conference of four of us to try to state the major problems confronting humanity. As if any four could succeed! Four of us gathered, and for some reason, in a willing suspension of disbelief, gave ourselves to the task. One of us was N. Scott Momaday, a

Pulitzer Prize–winning Kiowa, six foot seven, 250 pounds, bass voice. He paced: "The most important problem confronting mankind is to reinvent the sacred!" I, a biological scientist and M.D., was stunned. He could not speak that way! In about fifteen seconds I realized he was right. The four of us spoke and wrote a position statement that a global civilization of some form was emerging, that we could expect civilizational conflicts as our thirty or more civilizations crashed together, and that we needed a transnational mythic structure to undergird the emerging global civilization. I later wrote *Reinventing the Sacred* (2008), titled with credit to Momaday, seeking one sense of a natural God in the natural creativity of the universe in which we participate. God enough for me, I thought then and now. I hoped and hope that this is one source of spirituality for us and now it grows out of all of Part I of this book. Emergent life and us. But given Part II, if quantum measurement is part of a panpsychism as Penrose and Hameroff (Penrose 1989, 1994) and separately I, want to suggest, then aspects of the entire universe know and nonrandomly act at each measurement among independent or entangled quantum variables. If this arises among entangled quantum variables, they may "jointly know and decide." We do not know if there is some whispering form of Cosmic Mind playing a role in the becoming of the universe, perhaps among diverse and changing sets of entangled variables, Indra's Web? And if free will is at the foundations, or in us, no laws can entail the becoming of the universe. Sorry, Weinberg, but it is possible, and you with Newton denude us of our humanity. Why should we accept? Spirituality is said by Keats and the Romantic poets, Dylan Thomas, Shakespeare, Mozart's "Requiem," and by you walking the hills, smelling what no law entailed. "There is no truth beyond magic," wrote William Gaddis in *The Recognitions* (1955).

Fifth, we have progressively lost contact with Nature, teeming in teeming cities denuded of the Tangled Bank; Nature, which is our home and we are *of* it, not above it. We despoil the planet, wresting from Nature our due, as we in the Anthropocene are probably driving one of the most massive extinction events since the origin of life, destroying the accumulated co-created, interlaced functional wisdom of 3.7 billions of years of unstable becoming in our arrogance. What if all life is sentient? It just may be, given a possible panpsychism. In any case, elephants and whales are sentient, and we are killing them needlessly. We rape our home and ourselves in the name of gross national product, as price tags. Worse, we think the despoiled Earth and ecosystems can recover! We have not, as noted, succeeded in repopulating the overfished Grand Banks of Newfoundland. We kill soil species which lay the functional foundations for our lives in their rich functional couplings. Once lost, what depended upon them cannot be reconstituted. Don't we get it? What makes us think that, destroyed, that interwoven functionality can be re-created? We can no longer in the United States build the greatest of our steamships, The USS (The United States). We have lost the technologies and expertise. The Tasmanians, who arrived from Australia, forgot how to fish the rich seas. Once capacities, functionalities, are lost, they cannot be easily or ever rebuilt. We crash our planet and its living foundations, thinking we or "it" can repair, if ever the time comes and we actually care. Can we? Of Nature, we are, not above it, not ours to wrest our due as we unleash thousands of new chemical species into the atmosphere, waters, and land that is our home. How arrogant can we be? And in that arrogance, we no longer walk with Dylan Thomas on his October birthday. How very blind.

Sixth and new, we have not recognized the enormous influence of the ever-larger Adjacent Possibles we knowingly, and also with no intent, co-create and are almost ineluctably drawn into. History is not just a becoming; it is a becoming into what is *now possible*. That *Possible*, the Adjacent *Possible*, grew slowly 50,000 years ago, but it explodes now, thanks to purposeless teleology (Nagel 2012) and the antientropic processes of which I have written. We flow into the Adjacent Possible we co-create; we are "sucked" into its opportunities. We use these for better or worse. The banks too big to fail, in part hoping to spread risk, invented derivative financial instruments making no mops and mopping no floors, including mortgage-backed securities that offloaded responsibility of the local bank for repayment of its loan, thus incentivizing massive sale of risky mortgages, insured by credit default swaps that were not legally required to carry reserves to cover the losses, that led to the economic crash of 2008. But these financial instruments were themselves a flow into the Adjacent Possible enabled by the buying and selling of stocks in companies, itself enabled by a member of the Dutch East India Company breaking its rules and selling his shares to an outsider, establishing soon thereafter trade in stocks. Our massive international corporations, created legal individuals by law in the past century, adapt into the regulatory environment they lobby to mold to their advantage. That evolution is not prestatable in detail, for we do not know the loopholes that will be found. But once found and taken, the new actual behaviors and strategies open yet new Adjacent Possibles for new laws that try to control but have more loopholes, and actions that can be complements in unforeseen ways of unforeseen other new strategies. We flood ever more into the Adjacent Possibilities we co-create, not knowing what we are creating. This is a major aspect of cultural and economic evolution. Then a profound new issue: How do we garden the Adjacent Possibles we unknowingly in part co-create? What forms of governance and social weavings, beyond mere attempts at controlling the uncontrollable, must we conceive and put in place? What forms of bottom-up enablement coordinated across the scales of our individual and social lives in what ways?

Seventh: A vision of an emergent web of our thirty or more civilizations, able to respect the sacred roots of each, yet able to co-create as well. Perhaps this is my dream, a dream of ever-richer ways of being human. But how and what is wise, and how without killing one another in our fundamentalisms?

Our task is to conceive of a new transnational mythic structure beyond that of modernity. It cannot predict what will become; that cannot be prestated. But such a mythic structure can guide us. I hope that the seven issues just raised point us in some of the directions we must take to co-create that new mythic structure that just might take us beyond modernity.

Balancing Our Power Structures

Once the agricultural revolution set in, power structures emerged in the earliest civilizations in Mesopotamia, Egypt, and later China. Today, in the early twenty-first century, our power structures overwhelm us. In the United States, Congress, often well meaning, is also lobbied, often by past Congress members who are approved by current members of Congress to lobby Congress and have too-easy access to power, in the names of massive corporate interests. In the wake of the 2008 financial crisis brought on by banks too big to fail, the advent of "mortgage-backed securities" and "credit default swap," insurance on failure of mortgage-backed securities, for which the agencies selling the credit default swaps needed to carry no reserves to cover the insured loses, we have fewer banks even more too big to fail, and *no one* has gone to jail. TARP bailed them out, and perhaps it was necessary given the crisis they themselves created. As of 2015, efforts to regulate these banks are substantially corrupted by the lobbyists for the same banks who seek to modify the Dodd-Frank Act to leave loopholes for the benefit of the same banks. The US Supreme Court, counting corporations as "legal persons," passed "Citizens United" in the name of the First Amendment right to free speech and giving money as "speech," allowing vast sums of corporate monies (speech!) to flow into and corrupt our electoral processes.

We know all this, but we are doing nothing about it. Why not, and what can we do?

What Are Our Power Structures, How Do They Evolve, and Can They Be Tamed?

Remember, the Soviet Union collapsed. Apartheid is gone. Change is possible.

If we are to limit our power structures, should we so choose, we must understand them better. I trust it is acceptable to take a moment to sketch a simple model that may be helpful and can be developed. For the first issue is: how did we get here? I have discussed organisms as Kantian wholes that can adapt often unprestatably, into the very Adjacent Possibles evolution creates. In more powerful terms, evolution is "sucked" into the very opportunities it creates.

But the same is true of our power structures. A corporation, now a legal person, is also a Kantian functional whole living in its world. Its human members adopt roles that often fulfill their own joy, wealth, and power needs, and those of the organizations of which they are members. Major corporations and other organizations, from labor movements, to baseball teams, to political parties, are functional wholes with collective aims and collective functions that allow the organization to persist in its world. Or fail. In the meantime organizations create often unprestatable Adjacent Possibles, knowing not what they unleash, what they enable, and then are almost ineluctably drawn into the very opportunities, the unfolding in the Adjacent Possibles they and we co-create, and adapting to persist as power structures in unprestatable ways! Eisenhower's "military industrial complex" has metastasized into the NSA spying on all of us.

How do we understand this evolution of organizations, including power structures, into the ever-growing Adjacent Possible?

We need, among other things, a better understanding of organizations as adapting functional wholes. In Chapter 5, I suggested, as a starting point, an oversimple model to think about the evolution of our laws, and strategies or roles, and, with more work, organizations formed by such laws and strategies or roles and how such organizations evolve as Kantian wholes. The model consists in three kinds of "nodes"; squares represent laws, circles represent actions, triangles represent agents and motives. Each kind of node can have one or more inputs from each kind of node. In the simplest start of such modeling, each node is just on or off, a binary variable, so is "governed" by a logical rule called a Boolean function, which lists for all combinations of the inputs to that node, all combinations of on or off, for those inputs, if that node is on or off. For example, a given action may require two laws to be active and three motives, and to function on all five inputs, all must be present for the action to occur. Or one law may enable but a second block the action and two of the three motives enable, and the third block the action.

In the simplest start, time is also discrete, $t, t + 1$, and all nodes "update" synchronously, which is silly but useful. This is just a Boolean network, and if there are a total of n triangles plus circles plus squares, 2 is raised to the nth power combinations of the on-off states of these n variables. From each state the system flows to a next state and into an attractor, which may be a steady state.

Steady states are already interesting for they are self-instantiating, self-consistent forms of laws, actions, and motivations.

Such a system may have more than one steady state, each a different self-consistent set of laws and actions, and, formally, different "attractors" in the dynamics of this law action motive Boolean network. Each attractor may be reached by different initial states along transients that flow to that attractor, and the set of such transients is the "basin of attraction" of that attractor.

This oversimple model is also a beginning of a model of an organization that is at least a set of laws and actions, rules and roles, which is largely self-consistent. Actions, including roles in an organization, and organizational actions, are driven by "motives," deriving from emotions and beliefs of its people. These together form functional wholes that can be sufficient in their worlds of other organizations, the public, general laws, markets, or whatever else is in the environment.

To augment this oversimple model, one would want to include how individual actions converge on organizational decisions and actions, and how changing motives (including changes in value systems) altered the behavior of the system.

The total system is a dynamical system with attractors and now, with motives, a functional Kantian whole in its world.

The Evolution of Such Systems

But laws have loopholes that enable unforeseen new actions, which then require unforeseen new laws, driven by known and new motives, so the web of laws, actions, and motives evolves new squares, circles, and triangles with new Boolean logics in some way. This evolution is what creates the Adjacent Possible and is then the flow into that Adjacent Possible, driven by needs of the organizations and their members, including joy, greed, power, and cultural values and other values. These show up in the motives of the individuals and need to be added as part to the model of how the organization makes decisions and chooses actions.

We do not have such a body of theory yet, but something like it might be open to development, based just on toy theory and on data from real organizations as they evolve. As a hunch, we could study the evolution of myriad such systems, which can be defined classes, or "ensembles" of such systems to seek their typical, generic properties. We must take an ensemble approach for we cannot algorithmically prestate how such a system will find new opportunities in laws, roles, and motives. But we can study diverse classes or ensembles of such systems, their generic behaviors, and then try to map the real world.

Were we to have such a theory, rather like the RAFs in Chapter 14 on the origin of life, we could see how to alter these organizations. In RAFs, which achieve functional closure or sufficiency in their worlds, an irreducible RAF can be eliminated by deletion of one or a few chemicals. New RAFs can be spawned by transient exposure to a new chemical. Here is a beginning, all prestated, so inadequate to the real world, of a framework to think about our organizations and their evolution into their adjacent, often unprestatable, possibles, for example by finding loopholes in laws.

In fact, as noted in Chapter 15, the economy is also a RAF, with input goods the analogs of substrates, output goods the analogue of products, production functions the analogue of reactions, and tools mediating the production functions the analogue of catalysts. All are driven by motives.

The RAFs, whether about the origin of life, the economy, or generalized to organizations, are about "things" and transformations of "things" that may be abetted or hindered by the things themselves. What the things and transformations may, in fact, be is largely irrelevant. We are building theories of organization of process per se and their evolution. So, as these are independent of the stuff, these are not efficient cause laws, as Newton taught us. I want to think of these theories as affording "formal cause laws" when regularities, like criticality, can be found in the models and real world. Perhaps our organizations form a critical web.

An entire new way of thinking of the world, part of the theory of complexity, was born three decades ago and is growing.

We need such theories for we have co-created a modernity that only partly serves us. If we wish to change it, we need to know both the critical trigger functional elements whose alteration may yield change, with the caveat that we cannot foresee what we unleash when we do so. Because we cannot foresee, we must evolve our thinking. A simple example: In the United States, we sold guns and snowmobiles to the Inuit. They had used bows and dog sleds for centuries. Guns and snowmobiles were easier to use and adopted, but they were not produced by the Inuit. We quite deformed and destroyed a "functionally self-sufficient" way of life that had lasted perhaps thousands of years. Radical transformation is indeed possible. The Soviet Union and apartheid are really gone. What functional factors led to the collapse of the Soviet Union, beyond two obvious ones? We really do not want to live Marx's dream, "From each according to his capacity, to each according to his need." Marx thought human nature indefinitely malleable. Well, no. Second, a planned economy supposes it can prestate the market, rather like competitive general equilibrium. No. Life will find a new way.

If we wish to curb and mold our power structures, we need to know what log to pull out of the log jam. Then we need a means to do so, a far harder issue as power is concentrated away from we the people.

The concentration of power away from we the people begins to suggest an amended form of governance. We the people need sufficient structure to impinge *intricately* upon our power structures, sufficiently diverse in our *own* structure and interwoven with our power structures so we can affect them, to ourselves adapt in unprestatable ways and rapidly, so the power structures are playing "catch up," as the power structures and we co-evolve and co-create an unprestatable Adjacent Possible into which we all become, but their power is limited in ways now so very co-opted.

We do not know how to do this, but it suggests bottom-up enablement, means to coordinate across scales and interests, not easy at all, and dispersed adaptive forms of action that adapts in unprestatable adaptive ways. If this is sensible, or partially sensible, we need new social forms of organization able to mediate something like this. The *web* is just one means we the people might learn to use to form variegated, interacting social movements that can rapidly adapt and mobilize with enough people power to pull the right levers, not to destroy but to corral our power structures. We must adapt faster than they and in unprestatable ways. The military has the concept of the decision loop, the OODA loop. If we the people feel disempowered, and are, we do have means at our disposal, not to destroy but to garden more wisely inside the OODA loops of our power structures.

And we must evolve our values. What is our dream? How do we enable it when reason is an insufficient guide?

Civilizational Change After the Axial Age

The Axial Age is the fulcrum of much of the rest of history. Arnold Toynbee (1934–1961) wrote of the birth, maturation, and death of civilizations. He claims, perhaps too strongly, that the death of a civilization is typically accompanied by a spiritual rebirth that sets the frame for the new civilization. We can see evidence. The Greco-Roman civilization of antiquity gave way to the Christian world West and East that utterly transformed our values. The discovery in the thirteenth century of Lucretius' *De Rerum Natura* unleashed the Renaissance in Italy, with Michelangelo, da Vinci, and the Medici banking power structure. Again our value system vaulted. A new civilization emerged out of the Dark Ages; the Church view of humanity became the Mona Lisa. Then came Newton and the rise of science, and soon thereafter, the Age of Reason with the Enlightenment's scientific urge: down with the clerics, up with science for the betterment of mankind as we wrest our due from Nature. And also came its political ideals, writ into the US Constitution. Then we went to the Industrial Revolution, capitalism, communism, socialism, widening democracy, and modernity.

Quo Vadis?

Our diversity is co-created: Alicia Juarraro, in *Dynamics in Action* (2002), asks: "Could you cash a check 40,000 years ago?" Of course not. Think of the social, cultural, and legal inventions that had to occur to allow the new human activity of cashing a check. We *create* the Adjacent Possible cultural world that can *diversify* our ways of being human—yet preserve our cultural roots that are sacred to our diverse cultures as we co-mingle.

Thus, we want *not* a single culture, but a polyarchic diverse set of co-evolving and diversifying cultures as a vision beyond modernity.

Again, bottom-up enabled creativity and organization and governance, transparent by inhibiting by wise new laws ever-alterable, common law-like and some legislative, multiscale again, with competition between and collaboration among individuals and small and larger groups not only to attain prestated goals, but to *discover* new goals that we could not foresee. Wise revising of the Adjacent Possibles we co-create, only partially prestatable in what we enable, so ever adapting and also ever self-consciously revising the opportunities we are creating.

We do not want world government, too limiting, too dangerous in power accumulation.

We do want a diversity of civilizations co-creating new human cultural forms, respecting both appropriate aspects of our evolutionary past, our connection to nature, and those roots of our civilizations that are too sacred each to be modified without causing that civilization to collapse.

We do want our co-creativity; in all fields, whether painting or creating a business, it typically cannot be prestated. We solve the frame problem all the time and find deep joy in it. The flow into a tunable Adjacent Possible ever richer, but *not too fast*, permits us to enhance the diversity of our activities, how to be human. Again, Juarraro (2002): could we cash a check 40,000 years ago? Thus, the very fact that the abiotic universe as earlier chapter creates and invades ever new Adjacent Possibles, as does the evolving biosphere, economy, culture, says that our humanity too is part of a creative universe. There is no truth beyond magic: How much magic do we want? What we can wisely create together.

This is a call for a new Axial Age that reclaims enchantment and seeks new forms of the transcendence modernity has largely lost. But transformation of values is possible. Christianity rendered Roman values largely irrelevant. The Renaissance rendered the 1,000-year-old Church view of what our humanity is, and its values, largely obsolete. Newton was the start with Descartes of our dehumanization with res extensa, primary qualities, an entailed unfolding of all that can become, so no emergence, no "magic." We now can see that this modern science view is inadequate; we *are* part of a creative universe. Our deepest humanity is ours to seize. We must only begin to see it. What of modernity will we render irrelevant as we co-create?

We will not hurry to transform; the present is very entrenched in our lives. But we may, just may, transform, I hope gently. Revolutions kill people.

We profoundly need an overarching mythic structure to sustain the emerging global interwoven civilizations. At this hinge of history we need a new Axial Age.

Among the themes of that mythic structure beyond that of modernity are as follows:

1. Beyond Smith and Darwin as support for selfishness as the best or dominant driving force in the self-organization of society. The biosphere, from 150 bacterial species or more in your gut, richly functionally interwoven to achieve functional sufficiency *collaborate*, as well as compete, in ways we do not understand well. So do the myriad species in the biosphere functionally collaborate, even in multispecies autocatalytic systems (Ulanowitz 2009). So does our economy with its complements and substitutes. So do our

social organizations, all co-creating with one another into the unprestatable Adjacent Possibles we create.

2. Beyond overreliance on reason. We live into the future without knowing even what *can* happen and have for thousands of years. The Adjacent Possible explodes ever faster. We must live forward into it. Trust our knowing beyond reason. We have ever lived so and must moreso now. Life is not a known problem to be solved, but a becoming to be well discovered. Here is magic. And "There is no truth beyond magic" (Gaddis 1955).

3. A centrality for our rich emotional mature lives as individuals and in our social structures the dominant source of motivations and values and the pursuit of interwoven Happiness of which Jefferson wrote, but no each for each, but among us, beyond wealth where enough is enough. Not for bread alone. John Donne was right: No man is an island.

4. Spirituality as central to our lives and fulfillment. We have been spiritual, shaman onward, for 100,000 years or more. How dare our overscientistic, technological, First World civilization be so stupid as to deny this flood waiting again to flow? I wrote *Reinventing the Sacred* as one voice to say, of a natural but emergent biosphere beyond entailing law, here is one sense of God enough for me. But yet more if the universe is conscious and choosing and we with it. How dare we say *no* in our arrogance?

5. Rapprochement with Nature, no longer ours to wrest our due, we *of* Nature and a planet not to rape. No law entails that alive Nature, with species evolved for 3.7 billion years living an embodied wisdom garnered down these eons. And we kill the biosphere in this Anthropocene, causing a mass extinction in which that accumulated wisdom will vanish? How do we fail to notice that if too much of this living diversity vanishes, much more may collapse? We seem unable to re-create the fishing banks off the Grand Banks of Newfoundland that teemed with cod. Species can collectively form autocatalytic sets that can collapse (Ulanowitz 2009). Arrogant are we.

6. An ever-clearer understanding of how wisely to garden the Adjacent Possible into which we rush, but cannot prestate, with possible new forms of governance. We are not; we "become" and do so ever more without knowing what we co-create and are then sucked into beyond planning. Against design, we must "grow" ourselves.

7. We as creative in a creative universe, all a status nascendi of becoming, alive together, unfolding in an unknowable way. What is our wisdom?

What will we make of all this? We cannot prestate what we will do, but an enlarged mythic structure truly is ours to conceive and to guide us on how to live. I pray we may do so.

Conclusion

May this book be a contribution to our seeking anew our humanity in a creative universe in a global interwoven creative web of our many civilizations. Bless us.

References

Aho, Tommi, Juha Kessli, Stuart Kauffman, and Olli Yli-Harja. 2012. "Growth Efficiency as a Cellular Objective in *Escherichia coli*." *Physics ArXhiv* http://arxiv.org/abs/1203.4721

Arnhart, Larry. 2014. "The Evolution of Darwinian Liberalism." In *The Complexity of Liberty*, edited by Pablo Lucio Paredes and Sebastian Oleas. Quito, Ecuador: Universidad San Francisco de Quito.

Arrow, K., and G. Debreu. 1954. "Existence of an Equilibrium for a Competitive Economy." *Econometrica*, July: 265–290.

Ballivet, Marc, and Stuart A. Kauffman. 1987. French Patent Office. "Procede d'obtention d'ADN, ARN, peptides, polypeptides, ou proteines par une technique de recombinaison d'ADN." French Patent number 863683, issued December 24, 1987 and registered as 2,579,518.

Ballivet, Marc, and Stuart A. Kauffman. 1989. "Method of obtaining DNA, RNA, peptides, polypeptides or proteins by means of a DNA recombination technique." English Patent number 2183661, issued June 28, 1989.

Ballivet, Marc, and Stuart A. Kauffman. 1990. "Verfahren zur herstellung von peptiden polypeptiden oder proteinen." German Patent number 3,590,766.5–41, issued August 13, 1990.

Barad, Karen. 2007. *Meeting the Universe Halfway: Quantum Physics and the Entanglement of Matter and Meaning*. Durham, NC: Duke University Press.

Barrow, J. D., H. Sandvik, and J. Magueijo. 2002. "The Behaviour of Varying-Alpha Cosmologies." *Physical Review D* 65: 063504. doi:10.1103/PhysRevD.65.063504.

Bellah, R., and H. Joas. Eds. 2012. *The Axial Age and Its Consequences*. Cambridge, MA: Belknap Press of Harvard University Press.

Berkley, G. 1948–1957. *The Works of George Berkeley, Bishop of Cloyne*, edited by A. Al Luce and T. E. Jessop, 9 vols. London: Thomas Nelson and Sons.

Bohm, David, and Basil J. Hiley. 1993. *The Undivided Universe: An Ontological Interpretation of Quantum Theory*. London: Routledge & Kegan Paul.

Bohr, Neils. 1948. "On the Notions of Complementarity and Causality." *Dialectica* 2: 312–319.

Boltzmann, Ludwig. (1886) 1974. "The Second Law of Thermodynamics. Populare Schriften, Essay 3, Address to a Formal Meeting of the Imperial Academy of Science, 29 May 1886." In *Theoretical Physics and Philosophical Problems*, translated by S. G. Brush. Boston: Reidel.

Brand, Stewart. 1995. *How Buildings Learn: What Happens To Them When Built.* New York: Penguin.

Bray, Dennis. 2009. *Wetware: A Computer in Every Living Cell.* New Haven, CT: Yale University Press.

Chaitin, Gregory, Francisco A. Doria, and Newton da Costa. 2012. *Godel's Way: Exploits into an Undecidable World.* New York: CRC Press.

Chalmers, David J. 1996. *The Conscious Mind: In Search of a Fundamental Theory.* New York: Oxford University Press.

Chialvo, Dante R. 2012. "Critical Brain Dynamics at Large Scale." In *Criticality in Neural Systems*, edited by Ernst Niebur, Dietmar Plenz, and Heinz G. Schuster. New York: Wiley.

Chiarabelli, C., et al. 2006. *Chemistry and Biodiversity* 3: 827–837.

Conway, John H., and Simon Kochen. 2006. "The Strong Free Will Theorem." *Foundations of Physics* 36 (10): 1441–1473.

Crick, Francis. 1995. *The Astonishing Hypothesis: The Search for the Soul.* New York: Scribner.

Darwin, Charles. 1872. *The Expression of Emotion in Man and Animals.* London: John Murray.

Davies, Paul. 2006. *The Goldilocks Enigma: Why Is the Universe Just Right for Life?* New York: Houghten Mifflin.

Dawkins, Richard. 1976. *The Selfish Gene.* Oxford: Oxford University Press.

Daemer, David W., and Gail L. Barchfeld. 1982. Encapsulation of macromolecules by lipid vesicles under simulated prebiotic conditions. *Journal of Molecular Evolution* 18: 203.

Dehaene, Stanislas. 2014. *Consciousness and the Brain.* New York: Viking Press.

Descartes, Rene. 2014. "Discourse on the Method of Rightly Conducting the Reason and seeking the Truth in the Sciences." eBooks@Adelaide, 2014 (first published 1637).

DiVincenzo, David P., and Peter W. Shor. 1996. "Fault-Tolerant Error Correction with Efficient Quantum Codes." *Physics Review Letters* 77: 3260.

Dennett, Daniel. 1995. *Darwin's Dangerous Idea: Evolution and the Meaning of Life.* New York: Simon and Schuster.

Dennett, Daniel. 2003. *Freedom Evolves.* New York: Penguin.

Devins, Caryn, R. Koppl, and Stuart Kauffman. 2016. "Against Design." *Arizona State Law Journal* (in press).

Einstein, Albert, B. Podolsky, and N. Rosen. 1935. "Can Quantum-Mechanical Description of Physical Reality Be Considered Complete?" *Physics Review* 47: 777.

Eigen, Manfred. 1971. "The Self Organization of Matter and the Evolution of Biological Macromolecules." *Naturwissenschaften* 58: 465.

Eigen, Manfred, and P. Schuster. 1978. "Part A: Emergence of the Hypercycle." *Naturwissenschaften* 65: 7–41.

Emerson, Ralph Waldo. 2000. "Self Reliance." In *Transcendentalism, A Reader*, edited by Joel Myerson. New York: Oxford University Press.

Engel, Gregory S., Tessa R. Calhoun, Elizabeth L. Read, Tae-Kyu Ahn, Tomáš Mancal, Yuan-Chung Cheng, Robert E. Blankenship, and Graham R. Fleming. 2007. "Evidence for Wavelike Energy Transfer Through Quantum Coherence in Photosynthetic Systems." *Nature* 446 (7137): 782–786.

Epperson, Michael. *Quantum Mechanics and the Philosophy of Alfred North Whitehead*. Fordham University Press (2004).

Farmer, J. D., Stuart A. Kauffman, and N. H. Packard. 1986. "Autocatalytic Replication of Polymers." *Physica D* 2: 50–67.

Fernando, Chrisantha, Vera Vasas, Mauro Santos, Stuart Kauffman, and Eörs Szathmary. 2012. "Spontaneous Formation and Evolution of Autocatalytic Sets within Compartments." *Biology Direct* 7: 1.

Feynman, R., R. Leighton, and L. Sands. (1963) 2006, 2010. *The Feynman Lectures on Physics*. The Millennium Edition. Pasadena, CA: California Institute of Technology.

Gabora, Ll., and D. Aerts. 2002. "Contextualizing Concepts." Proceeding of the 15th International FLAIRS Conference (Special Track Categorization and Concept Representation: Models and Implications), Pensacola Beach, FL, May 14–17, American Association for Artificial Intelligence.

Gaddis, William. 1955. *The Recognitions*. New York: Harcourt Brace.

Gane, Simon, Dimitris Georganakis, Klio Maniati, Manolis Vamvakias, Nikitas Ragoussis, Efthimios M. C. Skoulakis, and Luca Turin. 2013. "Molecular Vibration-Sensing Component in Human Olfaction." *PLoS ONE* 8: e55780. doi:10.1371/journal.pone.0055780

Ganti, Timor. 2003. *The Principles of Life*. Oxford: Oxford University Press.

Gauger, Erik M., Elisabeth Rieper, John J. L. Morton, Simon C. Benjamin, and Vlatko Vedral. 2011. "Sustained Quantum Coherence and Entanglement in the Avian Compass." *Physics Review Letters* 106 (4): 040503.

Gillespie, Daniel T. 1976. "A General Method for Numerically Simulating the Stochastic Time Evolution of Coupled Chemical Reactions." *Journal of Computational Physics* 22 (4): 403–434. doi:10.1016/0021-9991(76)90041-3.

Gröblacher, S., T. Paterek, R. Kaltenbaek, S. Brukner, M. Zdotukoswki, M. Aspelmeyer, and A. Zeilinger. 2007. "An Experimental Test of Non-local Realism." *Nature* 446: 871–875.

Habermas, Jürgen. 1971. *Knowledge and Human Interests*. Beacon Press, Boston, 1971.

Haldane, J. B. S. 1954. "The Origins of Life." *New Biology* 16: 12–27.

Hanel, Rudolf, Stuart A. Kauffman, and Stefan Thurner. 2007. "Towards a Physics of Evolution: Critical Diversity Dynamics at the Edges of Collapse and Bursts of Diversification." *Physical Review E: Statistical, Nonlinear, and Soft Matter Physics* 76: 036110.

Hawking, Stephen, W. 2003. *The Illustrated Theory of Everything*. Beverly Hills: New Millennium Press.

Heisenberg, Werner. 1958. *Physics and Philosophy*. George Allen and Unwin.

Hordijk, Wim. 2013. "Autocatalytic Sets: From the Origin of Life to the Economy." *BioScience* 63 (11): 887–881. doi:10.1525/bio.2013.63.11.6.

Hordijk, Wim, Hein, J. and Mike Steel. 2010. "Autocatalytic Sets and the Origin of Life." *Entropy* 12 (7): 1733–1742.

Hordijk, Wim, Stuart Kauffman, and Mike Steel. 2011. "Required Levels of Catalysis for Emergence of Autocatalytic Sets in Models of Chemical Reaction Systems." *International Journal of Molecular Sciences* 12 (5): 3085–3101.

Hordijk, Wim, and Mike Steel. 2004. "Detecting Autocatalytic, Self-Sustaining Sets in Chemical Reaction Systems." *Journal of Theoretical Biology* 227: 451–461.

Hordijk, Wim, Mike Steel, and Stuart Kauffman. 2012. "The Structure of Autocatalytic Sets: Evolvability, Enablement, and Emergence." *Acta Biotheoretica* 60 (4): 379–392.

Jacob, Francois, and Jacques Monod. 1963. "Genetic Repression, Allosteric Inhibition and Cellular Differentiation." In *Cytodifferentiation and Macromolecular Synthesis*, edited by M. Locke. New York: Academic Press.

Jacob, François. 1977. "Evolution and Tinkering." *Science New Series*, 196(4295): 1161–1166.

James, William. 1909. *A Pluralistic Universe*. Lincoln: University of Nebraska Press.

Juarraro, Alicia. 2002. *Dynamics in Action: Intentional Behavior as a Complex System*. Cambridge, MA: MIT Press.

Jung, Carl G. 1971. *Psychological Types. Bollingen Series XX, Vol. 6*. Princeton, NJ: Princeton University Press.

Kant, Immanuel. (1892) 1951. *Critique of Judgment*, translated by J. H. Bernard. New York: Hafner.

Kauffman, Stuart A. (1969). Metabolic Stability and Epigenesis in Randomly Constructed Genetic Nets. *Journal of Theoretical Biology* 22, 437–467.

Kauffman, Stuart A. (1971). Cellular Homeostasis, Epigenesis, and Replication in Randomly Aggregated Macromolecular Systems. *Journal of Cybernetics* 1, 71–96.

Kauffman, Stuart A. (1986). Autocatalytic Sets of Proteins. *Journal of Theoretical Biology* 119, 1–24.

Kauffman, Stuart. 1993. *Origins of Order: Self Organization and Selection in Evolution*. New York: Oxford University Press.

Kauffman, Stuart. 1995. *At Home in the Universe*. New York: Oxford University Press.

Kauffman, Stuart. 2000. *Investigations*. New York: Oxford University Press.

Kauffman, Stuart. 2008. *Reinventing the Sacred*. New York: Basic Books.

Kauffman, Stuart A. 2012. "Is The Possible Ontologically Real?" NPR13.7 Cosmos and Culture.

Kauffman, Stuart. 2014. "Beyond the Stalemate: Mind Body, Quantum Mechanics, Free Will, Possible Panpsychism, Possible Solution to the Quantum Enigma." Accessed October 1, 2015. http://arxiv.org/ftp/arxiv/papers/1410/1410.2127.pdf.

Kauffman, Stuart A. 2016. *Answering Descartes: Beyond Turing in The Once and Future Turing*, edited by S. Barry Cooper. Cambridge, UK: Cambridge University Press (in press).

Kauffman, Stuart A., and Marc Ballivet. 1998. "Method of identifying a stochastically-generated peptide, polypeptide, or protein having ligand binding property and compositions thereof." US Patent number 5,723,323, issued March 3, 1998.

Kauffman, Stuart, Gabor Vattay, and Samuli Niiranen. 2014. "Uses of systems of degrees of freedom poised between fully quantum and fully classical." US Patent 8,849,580,B2, issued September 30, 2014.

Kull, Kalavi. 2009. "Biosemiotics: 'To Know, What Life Knows.'" *Cybernetics and Human Knowing* 16: 81–88.

Labean, T. H., T. R. Butt, Stuart Kauffman, and E. A. Schultes. 2011. "Protein Folding Absent Selection." *Genes (Basel)* 2 (3): 608–626. doi: 10.3390/genes2030608.

Lambert, N., Y-N. Chen, Y-C Cheng, C-M Li, G-Y Chen, and F. Nori. 2013. "Quantum Biology." *Nature Physics* 9 10–18. doi:10.1038/nphys2474

Lee, David H., Kay Severin, Yohei Yokobayashi, and M. Reza Ghadiri. 1997. "Emergence of Symbiosis in Peptide Self-Replication through a Hypercyclic Network." *Nature* 390: 591.

Libet, Benjamin. 1985. "Unconscious Cerebral Initiative and the Role of Conscious Will in Voluntary Action." *Behavioral and Brain Sciences* 8 (4): 529–566.

Longo, Giuseppe, and Mael Montevil. 2014. *Perspectives on Organisms: Biological Time, Symmetries, and Singularities*. Berlin: Springer.

Longo, Giuseppe, Mael Montevil, and Stuart Kauffman. "No Entailing Laws, But Enablement in the Evolution of Life." In *Proceedings of the Fourteenth International Conference on Genetic and Evolutionary Computation Conference Companion*, pp. 1379–1392. doi:10.1145/2330784/2330946.

Maturana, Humberto, and Francisco Varela. (1980). *Autopoiesis and Cognition: The Realization of the Living*, edited by Robert S. Cohen and Marx W. Wartofsky. 2nd edition. Dordrecht, The Netherlands: Boston Studies in the Philosophy of Science.

McCulloch, Warren, and Walter Pitts. 1943. "The Logical Calculus of Ideas Immanent in Nervous Activity." *Bulletin of Mathematical Biophysics* 5: 115–133.

Merleau-Ponty, Maurice. 1969. *The Visible and the Invisible*. Evanston: Northwestern University Press.

Miller, Stanley L. 1953. "Production of Amino Acids Under Possible Primitive Earth Conditions." *Science* 117 (3046): 528–529.

Mossel, Elchanan, and Mike Steel. 2005. "Random biochemical networks and the probability of self-sustaining autocatalysis." *Journal of Theoretical Biology* 233 (3): 327–336.

Nagel, Thomas. 1974. "What Is It Like to Be a Bat?" *Philosophical Review* 83 (4): 1 435–450.

Nagel, Thomas. 2012. *Mind and Cosmos: Why the Materialism Neo-Darwinian Conception of Nature Is Almost Certainly False.* New York: Oxford University Press.

Nicholis, Gregoire, and I. Prigogine. 1977. *Self-Organization in Nonequilibrium Systems.* New York: John Wiley and Sons.

Nietzsche, Friedrich. 1968. "The Geneology of Morals." In *The Basic Writings of Nietzsche*, translated by Walter Kaufmann. New York: Modern Library Giants, Random House.

Odum, H. T. 1995. "Self Organization and Maximum Power." In *Maximum Power*, edited by C. A. S. Hall, pp. 311–364. Niwot: University Press of Colorado.

O'Reilly, Edward J., and Alexandra Olaya-Castro. 2014. "Non-Classicality of the Molecular Vibrations Assisting Exciton Energy Transfer at Room Temperature." *Nature Communications* 5, Article number 3012. doi:10.1038/ncomms4012

Orgel, Leslie E., and R. Lorhmann. 1974. *Accounts of Chemical Research* 7: 368–377.

Oparin, A. I. 1968. *The Origin and Development of Life* (NASA TTF-488). Washington, DC: D.C.L GPO.

Patil, Y. S., S. Chakram, and M. Vengaltore. 2015. "Measurement-Induced Localization of an Ultracold Lattice Gas." *Physical Review Letters* 115, 140402. arXiv:1411.2678.

Peil, Katherine T. 2014. "Emotion: The Self—Regulatory Sense." *Global Advances in Health and Medicine* 3 (2): 80–108.

Penrose, Roger. 1989. *The Emperor's New Mind.* New York: Oxford University Press.

Penrose, Roger. 1994. *Shadows of the Mind.* New York: Oxford University Press.

Penrose, Roger. 2004. *The Road to Reality.* London: Jonathon Cape.

Penrose, Roger, Abner Shimony, Nancy Cartwright, and Stephen Hawking. 1997. *The Large, the Small and the Human Mind.* Cambridge: Cambridge University Press.

Peirce, C. S. 1969. *Collected Papers Vols I and II*, edited by Charles Hartshorne and Paul Weiss. Cambridge: Harvard University Press.

Prigogine, Ilya, and G. Nicolis. 1977. *Self-Organization in Non-Equilibrium Systems.* New York: Wiley.

Pross, Addy. 2012. *What Is Life: How Chemistry Becomes Biology.* New York: Oxford University Press.

Quine, Willard V. 1951. "Two Dogmas of Empiricism." *Philosophical Review* 60 (1): 20–43.

Radin, Dean. 2006. *Entangled Minds.* New York: Paraview Pocket Books.

Radin, Dean, Leena Michel, Karla Galdamez, Paul Wendland, and Robert Rickenbach. 2012. "Consciousness and the Double-Slit Interference Pattern: Six Experiments." *Physics Essays* 25: 2.

Radin, Dean, Leena Michel, James Johnston, and Arnaud Delome. 2013. "Psychophysical Interactions with a Double-Slit Interference Pattern." *Physics Essays* 26: 4.

Rawls, John. 1971. *A Theory of Justice*. New York: Belknap.

Rebentrost, P., M. Mohseni, I. Kassal, S. Lloyd, and A. Aspuru-Guzik. 2009. "Environment-Assisted Quantum Transport." *New Journal of Physics*. doi:10.1088/1367-2630/11/3/033003.

Ringbauer, Martin, Ben Duffus, Cyril Branciard, Eric G. Cavalcanti, Andrew G. White, and Alessandro Fedrizzi. 2015. "Measurements on the Reality of the Wavefunction." *Nature Physics* 11: 249–254.

Rosen, Robert. 1991. *Life Itself*. New York: Columbia University Press.

Rosenblum, Bruce, and Fred Kuttner. 2006. *Quantum Enigma: Physics Encounters Consciousness*. New York: Oxford University Press.

Russell, Bertrand. 1956. "The Philosophy of Logical Atomism." In *Bertrand Russell, Logic and Knowledge Essays 1901–1950*, edited by Robert March. London: George Allen and Unwin.

Schrodinger, E. 1944. *What Is Life? Mind and Matter*. Cambridge: Cambridge University Press.

Shor, Peter. 1997. *Fault-Tolerant Quantum Computation*. Accessed October 2, 2015. http://arxiv.org/pdf/quant-ph/9605011.pdf

Smolin, Lee. 2013. *Time Reborn: From the Crisis in Physics to the Future of the Universe*. New York: Houghton Mifflin Harcourt.

Snow, Charles Percy. (1959) 2001. *The Two Cultures*. London: Cambridge University Press.

Stano, P., E. Wehrli, and P. L. Luisi. 2006. "Insights on the Oleate Vesicles Self-Reproduction." *Journal of Physics: Condensed Matter* 18: S2231–2238.

Stapp, Henry. 2011. *The Mindful Universe: Quantum Mechanics and the Participating Observer*. New York: Springer Verlag.

Suppe, Frederick. 1999. "The Positivist Model of Scientific Theories." In *Scientific Inquiry*, edited by Robert Klee, 16–24. New York: Oxford University Press.

Susskind, Leonard. 2006. *The Cosmic Landscape: String Theory and the Illusion of Intelligent Design*. New York: Little Brown.

Thoreau, Henry. 1854. *On Walden Pond*. Boston: Ticknor and Fields.

Tiwari, Vivek, William K. Peters, and David M. Jonas. 2013. "Electronic Resonance with Anticorrelated Pigment Vibrations Drives Photosynthetic Energy Transfer Outside the Adiabatic Framework." *Proceedings of the National Academy of Sciences USA* 110 (4): 1203–1208.

Toynbee, Arnold. 1934–1961. *A Study of History, Vols I—XII*. Oxford: Oxford University Press.

Tse, Peter Ulric. 2013. *The Neuronal Basis of Free Will: Criterial Causation*. Cambridge, MA: MIT Press.

Turing, Alan. 1953. "The Chemical Basis of Morphogenesis." *Philosophical Transactions of the Royal Society of London* B237: 37.

Turing, Alan. 1968. "Intelligent Machinery." In *Cybernetics: Key Papers*, edited by C. R. Evans and A. D. J. Robertson. Baltimore: University Park Press.

Ulanowitz, Robert. 2009. *The Third Window: A Natural Life Beyond Newton and Darwin.* New York: Templeton Press.

Vaidya, Nilesh, Michael L. Manapat, Irene A. Chen, Ramon Xulvi-Brunet, Eric J. Hayden, and Niles Lehman. 2012. "Spontaneous Network Formation among Cooperative RNA Replicators." *Nature* 491: 72–77. doi:10.1038/nature11549.

Vasas, Vera, Chrisantha Fernando, Mauro Santos, Stuart Kauffman, and Eörs Szathmary. 2012. "Evolution before Genes." *Biology Direct* 7: 1.

Vattay, Gabor, and Istvan Csabai. 2015. "Environment Assisted Quantum Transport in Organic Molecules." arxiv.org/abs/1503.00178

Vattay, Gabor, Stuart Kauffman, and Samuli Niiranen. 2012. "Quantum Biology on the Edge of Quantum Chaos." *PLoS One* 9 (3): e89017.

Vattay, Gabor, Dennis Salahub, Istvan Csabai, Ali Nassmi, and Stuart Kauffman. 2015. "Quantum Criticality at the Origin of Life." *Journal of Physics* 626: 012023.

Villani, Marco, Alessandro Filisetti, Alex Graudenzi, Chiara Damiani, Timoteo Carletti, and Roberto Serra. 2014. "Growth and Division in a Dynamic Protocell Model." *Life* 4: 837–864.

von Neumann, John. (1933) 1955. *Mathematical Foundations of Quantum Mechanics.* Princeton, NJ: Princeton University Press.

Weber, Andreas. 2013. *Enlivement: Towards a Fundamental Shift in the Concepts of Nature, Culture and Politics.* Berlin, Germany: Heinrich—Bol—Stiftung.

Wagner, N., and Gonen Ashkenasy. 2009. "Systems Chemistry: Logical Gates, Arithmetic Units and Network Motifs in Small Networks." *Chemistry—A European Journal* 15: 1765–1775.

Weinberg, Stephen. 1994. *Dreams of a Final Theory: The Scientist's Search for the Ultimate Laws of Nature.* New York: First Vintage Books.

Wheeler, John A. 1990. "Information, Physics, Quantum: The Search for Links." In *Complexity, Entropy, and the Physics of Information*, edited by W. Zurek. Redwood City, CA: Addison-Wesley.

Whitehead, Alfred North. 1978. *Process and Reality, Gifford Lectures 1927–1928*, edited by David Ray Griffin and Donald W. Sherburne. New York: The Free Press, Simon and Schuster.

Whitehead, Alfred North, and Bertrand Russell. 1925. *Principia Mathematica.* Cambridge: Cambridge University Press.

Wittgenstein, Ludwig. 1921. *Tractatus Logico-Philosphicus: Logisch-Philosophische Abhandlung*, edited by Wilhelm Ostwald. Annalen der Naturphilosophie, 14.

Wittgenstein, Ludwig. (1953) 2001. *Philosophical Investigations.* Oxford: Blackwell.

Yonath, Ada. 2014. Lecture at II International Symposium on Evolutionary Biology, Joao Pessoa, Paraiba—Brazil, December 8–12.

Zeh, Dieter. 2007, April 30. *FQXi.* "Ultimate Reality: Wave Function Collapse Demystified." [forum post]. http://fqxi.org/community/forum/topic/39.

Zurek, W. H. 1991. "Decoherence and the Transition from Quantum to Classical." *Physics Today* 44: 36.

Index

Printed in the USA/Agawam, MA
January 15, 2024

859406.046